THE ILLUSTRATED GUIDE TO EDIBLE WILD PLANTS

Department of The Army

Must Have Books
503 Deerfield Place
Victoria, BC
V9B 6G5
Canada

ISBN: 9781773238159

Copyright 2021 – Must Have Books

All rights reserved in accordance with international law. No part of this book may be reproduced or transmitted in any form or by any means, electronic or mechanical, including photocopying, recording, or by any information storage or retrieval system, except in the case of excerpts by a reviewer who may quote brief passages in an article or review, without written permission from the publisher.

TABLE OF CONTENTS

Part 1
Edible Plants 1
Plant Identification 2
Universal Edibility Test 5
Edible Plants... 7

Part 2
Poisonous Plants 119
Rules for Avoiding Poisonous Plants................... 120
Contact Dermatitis..................................... 120
Ingestion Poisoning 121
Poisonous Plants 122

PART 1

EDIBLE PLANTS

In a survival situation, plants can provide food and medicine. Their safe usage requires absolutely positive identification, knowing how to prepare them for eating, and knowing any dangerous properties they might have. Familiarity with botanical structures of plants and information on where they grow will make them easier to locate and identify.

Plant Identification

You identify plants, other than by memorizing particular varieties through familiarity, by using such factors as leaf shape and margin, leaf arrangements, and root structure.

The basic leaf margins (see Figure 1.1) are toothed, lobed, and toothless or smooth.

These leaves may be lance-shaped, elliptical, egg-shaped, oblong, wedge-shaped, triangular, long-pointed, or top-shaped (Figure 1.2).

The basic types of leaf arrangements (Figure 1.3) are opposite, alternate, compound, simple, and basal rosette.

The basic types of root structures (Figure 1.4) are the bulb, clove, taproot, tuber, rhizome, corm, and crown. Bulbs are familiar to us as onions and, when sliced in half, will show concentric rings. Cloves are those bulblike structures that remind us of garlic and will separate into small pieces when broken apart. This characteristic separates wild onions from wild garlic. Taproots resemble carrots and may be single-rooted or branched, but usually only one plant stalk arises from each root. Tubers are like potatoes and daylilies and you will find these structures either on strings or in clusters underneath the

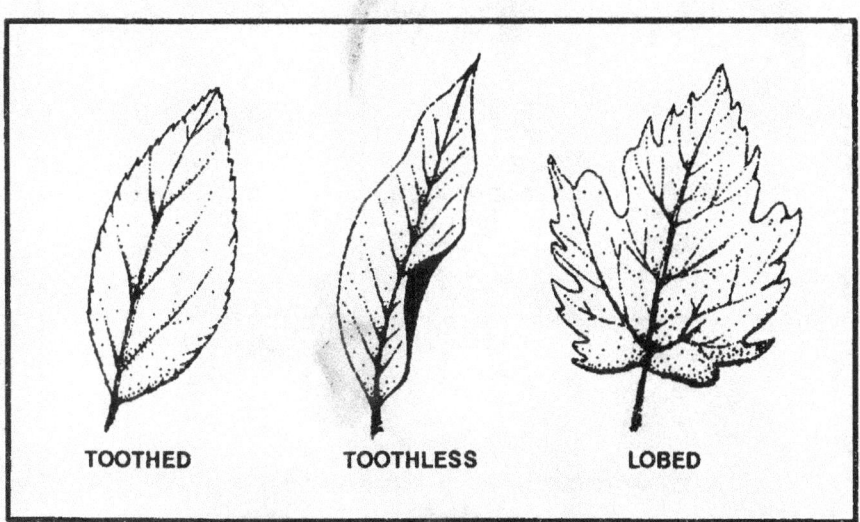

Figure 1-1. Leaf margins.

parent plants. Rhizomes are large creeping rootstocks or underground stems and many plants arise from the "eyes" of these roots. Corms are similar to bulbs but are solid when cut rather than possessing rings. A crown is the type of root structure found on plants such as asparagus and looks much like a mophead under the soil's surface.

Learn as much a possible about plants you intend to use for food and their unique characteristics. Some plants have both edible and poisonous parts. Many are edible only at certain times of the year. Others may have poisonous relatives that look very similar to the ones you can eat or use for medicine.

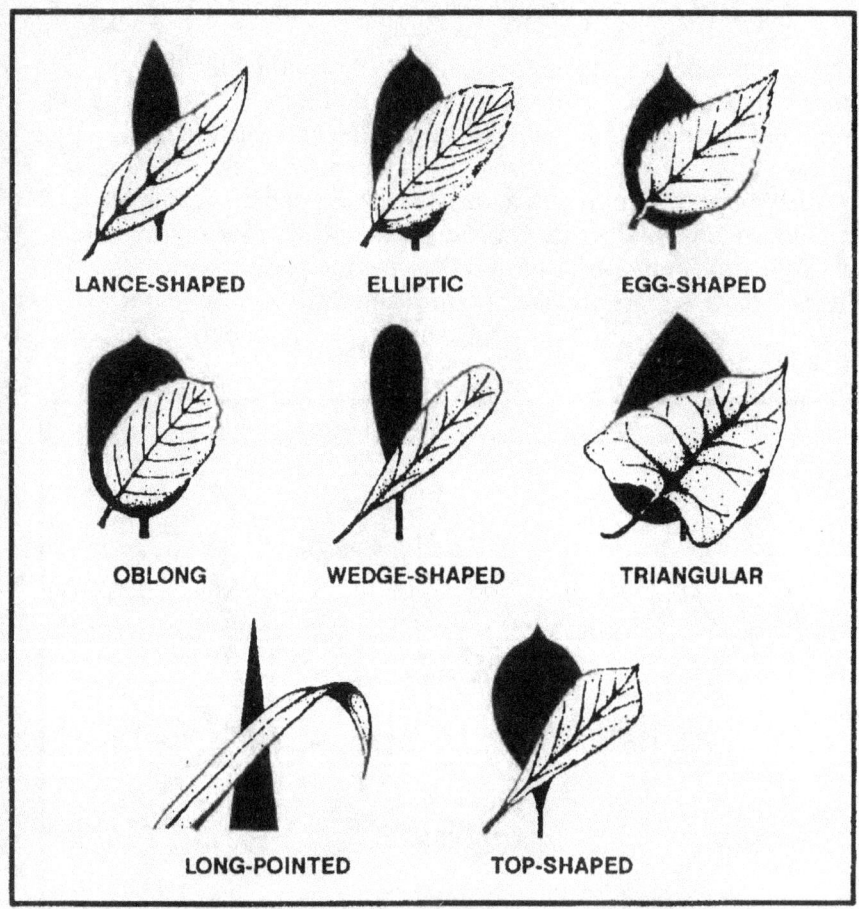

Figure 1-2. Leaf shapes.

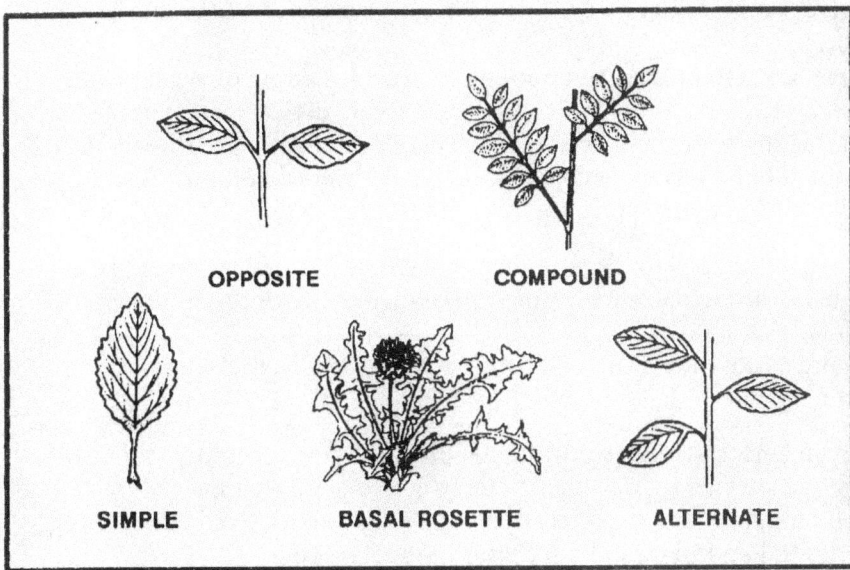

Figure 1-3. Leaf arrangements.

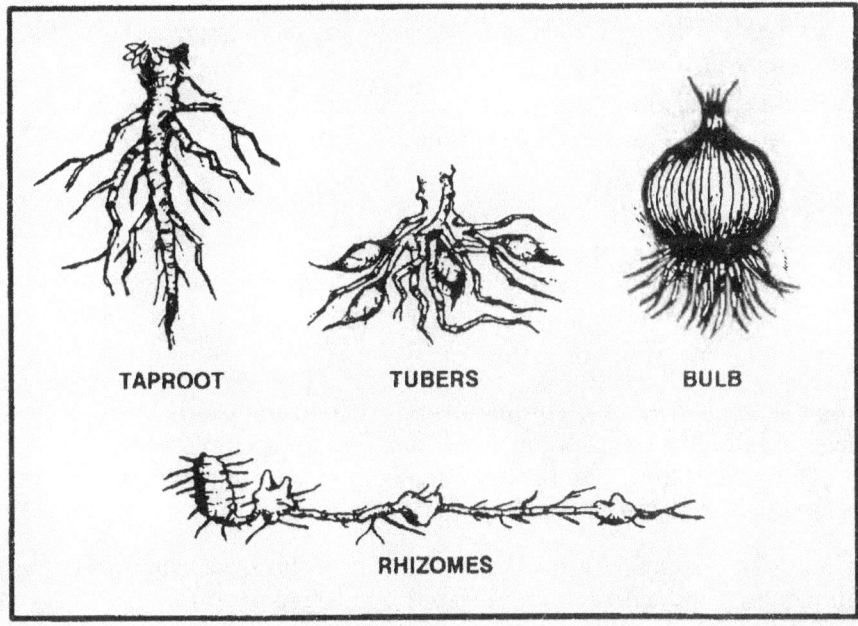

Figure 1-4. Root structures.

Universal Edibility Test

There are many plants throughout the world. Tasting or swallowing even a small portion of some can cause severe discomfort, extreme internal disorders, and even death. Therefore, if you have the slightest doubt about a plant's edibility, apply the Universal Edibility Test (Figure 1.5) before eating any portion of it.

Before testing a plant for edibility, make sure there are enough plants to make the testing worth your time and effort. Each part of a plant (roots, leaves, flowers, and so on) requires more than 24 hours to test. Do not waste time testing a plant that is not relatively abundant in the area.

Remember, eating large portions of plant food on an empty stomach may cause diarrhea, nausea, or cramps. Two good examples of this are such familiar foods as green apples and wild onions. Even after testing plant food and finding it safe, eat it in moderation.

You can see from the steps and time involved in testing for edibility just how important it is to be able to identify edible plants.

To avoid potentially poisonous plants, stay away from any wild or unknown plants that have—
- Milky or discolored sap.
- Beans, bulbs, or seeds inside pods.
- Bitter or soapy taste.
- Spines, fine hairs, or thorns.
- Dill, carrot, parsnip, or parsleylike foliage.
- "Almond" scent in woody parts and leaves.
- Grain heads with pink, purplish, or black spurs.
- Three-leaved growth pattern.

Using the above criteria as eliminators when choosing plants for the Universal Edibility Test will cause you to avoid some edible plants. More important, these criteria will often help you avoid plants that are potentially toxic to eat or touch.

Learn as much as possible about the plant life of the areas where you train regularly and where you expect to be traveling or working.

1	Test only one part of a potential food plant at a time.
2	Separate the plant into its basic components—leaves, stems, roots, buds, and flowers.
3	Smell the food for strong or acid odors. Remember, smell alone does not indicate a plant is edible or inedible.
4	Do not eat for 8 hours before starting the test.
5	During the 8 hours you abstain from eating, test for contact poisoning by placing a piece of the plant part you are testing on the inside of your elbow or wrist. Usually 15 minutes is enough time to allow for a reaction.
6	During the test period, take nothing by mouth except purified water and the plant part you are testing.
7	Select a small portion of a single part and prepare it the way you plan to eat it.
8	Before placing the prepared plant part in your mouth, touch a small portion (a pinch) to the outer surface of your lip to test for burning or itching.
9	If after 3 minutes there is no reaction on your lip, place the plant part on your tongue, holding it there for 15 minutes.
10	If there is no reaction, thoroughly chew a pinch and hold it in your mouth for 15 minutes. **Do not swallow.**
11	If no burning, itching, numbing, stinging, or other irritation occurs during the 15 minutes, swallow the food.
12	Wait 8 hours. If any ill effects occur during this period, induce vomiting and drink a lot of water.
13	If no ill effects occur, eat 0.25 cup of the same plant part prepared the same way. Wait another 8 hours. If no ill effects occur, the plant part as prepared is safe for eating.

CAUTION

Test all parts of the plant for edibility, as some plants have both edible and inedible parts. Do not assume that a part that proved edible when cooked is also edible when raw. Test the part raw to ensure edibility before eating raw. The same part or plant may produce varying reactions in different individuals.

Figure 1-5. Universal Edibility Test

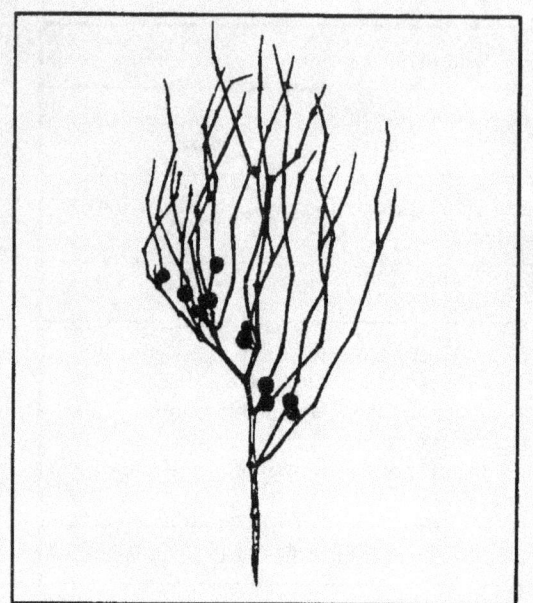

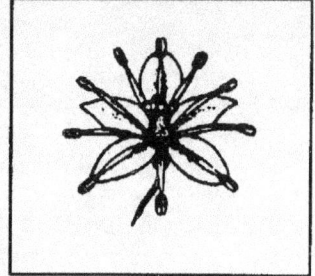

Abal
Calligonum comosum

Description: The abal is one of the few shrubby plants that exists in the shady deserts. This plant grows to about 1.2 meters, and its branches look like wisps from a broom. The stiff, green branches produce an abundance of flowers in the early spring months (March, April).

Habitat and Distribution: This plant is found in desert scrub and waste in any climatic zone. It inhabits much of the North African desert. It may also be found on the desert sands of the Middle East and as far eastward as the Rajputana desert of western India.

Edible Parts: This plant's general appearance would not indicate its usefulness to the survivor, but while this plant is flowering in the spring, its fresh flowers can be eaten. This plant is common in the areas where it is found. An analysis of the food value of this plant has shown it to be high in sugar and nitrogenous components.

Acacia
Acacia farnesiana

Description: Acacia is a spreading, usually short tree with spines and alternate compound leaves. Its individual leaflets are small. Its flowers are ball-shaped, bright yellow, and very fragrant. Its bark is a whitish-gray color. Its fruits are dark brown and podlike.

Habitat and Distribution: Acacia grows in open, sunny areas. It is found throughout all tropical regions.

Note: There are about 500 species of acacia. These plants are especially prevalent in Africa, southern Asia, and Australia, but many species are found in the warmer and drier parts of America.

Edible Parts: Its young leaves, flowers, and pods are edible raw or cooked.

Agave
Agave species

Description: These plants have large clusters of thick, fleshy leaves borne close to the ground and surrounding a central stalk. The plants flower only once, then die. They produce a massive flower stalk.

Habitat and Distribution: Agaves prefer dry, open areas. They are found throughout Central America, the Caribbean, and parts of the western deserts of the United States and Mexico.

Edible Parts: Its flowers and flower buds are edible. Boil them before eating.

CAUTION
The juice of some species causes dermatitis in some individuals.

Other Uses: Cut the huge flower stalk and collect the juice for drinking. Some species have very fibrous leaves. Pound the leaves and remove the fibers for weaving and making ropes. Most species have thick, sharp needles at the tips of the leaves. Use them for sewing or making hooks. The sap of some species contains a chemical that makes the sap suitable for use as a soap.

Almond
Prunus amygdalus

Description: The almond tree, which sometimes grows to 12.2 meters, looks like a peach tree. The fresh almond fruit resembles a gnarled, unripe peach and grows in clusters. The stone (the almond itself) is covered with a thick, dry, woolly skin.

Habitat and Distribution: Almonds are found in the scrub and thorn forests of the tropics, the evergreen scrub forests of temperate areas, and in desert scrub and waste in all climatic zones. The almond tree is also found in the semidesert areas of the Old World in southern Europe, the eastern Mediterranean, Iran, the Middle East, China, Madeira, the Azores, and the Canary Islands.

Edible Parts: The mature almond fruit splits open lengthwise down the side, exposing the ripe almond nut. You can easily get the dry kernel by simply cracking open the stone. Almond meats are rich in food value, like all nuts. Gather them in large quantities and shell them for further use as survival food. You could live solely on almonds for rather long periods. When you boil them, the kernel's outer covering comes off and only the white meat remains.

Amaranth
Amaranthus species

Description: These plants, which grow 90 centimeters to 150 centimeters tall, are abundant weeds in many parts of the world. All amaranth have alternate simple leaves. They may have some red color present on the stems. They bear minute, greenish flowers in dense clusters at the top of the plants. Their seeds may be brown or black in weedy species and light-colored in domestic species.

Habitat and Distribution: Look for amaranth along roadsides, in disturbed waste areas, or as weeds in crops throughout the world. Some amaranth species have been grown as a grain crop and a garden vegetable in various parts of the world, especially in South America.

Edible Parts: All parts are edible, but some may have sharp spines you should remove before eating. The young plants or the growing tips of older plants are an excellent vegetable. Simply boil the young plants or eat them raw. Their seeds are very nutritious. Shake the tops of the older plants to get the seeds. Eat the seeds raw, boiled, ground into flour, or popped like popcorn.

Arctic willow
Salix arctica

Description: The arctic willow is a shrub that never exceeds more than 60 centimeters in height and grows in clumps that form dense mats on the tundra.

Habitat and Distribution: The arctic willow is common on tundras in North America, Europe, and Asia. You can also find it in some mountainous areas in temperate regions.

Edible Parts: You can collect the succulent, tender young shoots of the arctic willow in early spring. Strip off the outer bark of the new shoots and eat the inner portion raw. You can also peel and eat raw the young underground shoots of any of the various kinds of arctic willow. Young willow leaves are one of the richest sources of vitamin C, containing 7 to 10 times more than an orange.

 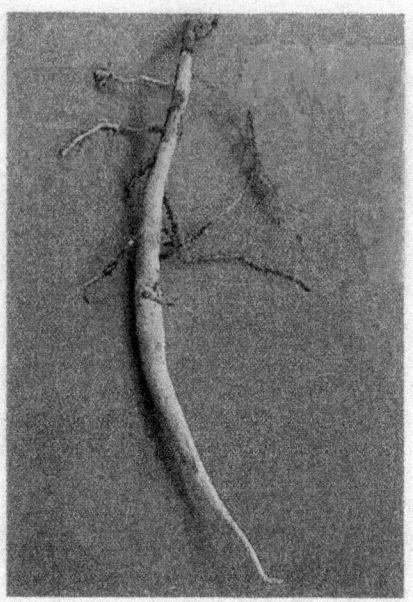

Arrowroot
Maranta and Sagittaria species

Description: The arrowroot is an aquatic plant with arrow-shaped leaves and potatolike tubers in the mud.

Habitat and Distribution: Arrowroot is found worldwide in temperate zones and the tropics. It is found in moist to wet habitats.

Edible Parts: The rootstock is a rich source of high quality starch. Boil the rootstock and eat it as a vegetable.

Asparagus
Asparagus officinalis

Description: The spring growth of this plant resembles a cluster of green fingers. The mature plant has fernlike, wispy foliage and red berries. Its flowers are small and greenish in color. Several species have sharp, thorn-like structures.

Habitat and Distribution: Asparagus is found worldwide in temperate areas. Look for it in fields, old homesites, and fencerows.

Edible Parts: Eat the young stems before leaves form. Steam or boil them for 10 to 15 minutes before eating. Raw asparagus may cause nausea or diarrhea. The fleshy roots are a good source of starch.

> **WARNING**
> Do not eat the fruits of any species since some are toxic.

Bael fruit
Aegle marmelos

Description: This is a tree that grows from 2.4 to 4.6 meters tall, with a dense spiny growth. The fruit is 5 to 10 centimeters in diameter, gray or yellowish, and full of seeds.

Habitat and Distribution: Bael fruit is found in rain forests and semievergreen seasonal forests of the tropics. It grows wild in India and Burma.

Edible Parts: The fruit, which ripens in December, is at its best when just turning ripe. The juice of the ripe fruit, diluted with water and mixed with a small amount of tamarind and sugar or honey, is sour but refreshing. Like other citrus fruits, it is rich in vitamin C.

Bamboo
Various species including *Bambusa, Dendrocalamus, Phyllostachys*

Description: Bamboos are woody grasses that grow up to 15 meters tall. The leaves are grasslike and the stems are the familiar bamboo used in furniture and fishing poles.

Habitat and Distribution: Look for bamboo in warm, moist regions in open or jungle country, in lowland, or on mountains. Bamboos are native to the Far East (Temperate and Tropical zones) but have been widely planted around the world.

Edible Parts: The young shoots of almost all species are edible raw or cooked. Raw shoots have a slightly bitter taste that is removed by boiling. To prepare, remove the tough protective sheath that is coated with tawny or red hairs. The seed grain of the flowering bamboo is also edible. Boil the seeds like rice or pulverize them, mix with water, and make into cakes.

Other Uses: Use the mature bamboo to build structures or to make containers, ladles, spoons, and various other cooking utensils. Also use bamboo to make tools and weapons. You can make a strong bow by splitting the bamboo and putting several pieces together.

CAUTION
Green bamboo may explode in a fire. Green bamboo has an internal membrane you must remove before using it as a food or water container.

Banana and plantain
Musa species

Description: These are treelike plants with several large leaves at the top. Their flowers are borne in dense hanging clusters.

Habitat and Distribution: Look for bananas and plantains in open fields or margins of forests where they are grown as a crop. They grow in the humid tropics.

Edible Parts: Their fruits are edible raw or cooked. They may be boiled or baked. You can boil their flowers and eat them like a vegetable. You can cook and eat the rootstocks and leaf sheaths of many species. The center or "heart" of the plant is edible year-round, cooked or raw.

Other Uses: You can use the layers of the lower third of the plants to cover coals to roast food. You can also use their stumps to get water. You can use their leaves to wrap other foods for cooking or storage.

Baobab
Adansonia digitata

Description: The baobab tree may grow as high as 18 meters and may have a trunk 9 meters in diameter. The tree has short, stubby branches and a gray, thick bark. Its leaves are compound and their segments are arranged like the palm of a hand. Its flowers, which are white and several centimeters across, hang from the higher branches. Its fruit is shaped like a football, measures up to 45 centimeters long, and is covered with short dense hair.

Habitat and Distribution: These trees grow in savannas. They are found in Africa, in parts of Australia, and on the island of Madagascar.

Edible Parts: You can use the young leaves as a soup vegetable. The tender root of the young baobab tree is edible. The pulp and seeds of the fruit are also edible. Use one handful of pulp to about one cup of water for a refreshing drink. To obtain flour, roast the seeds, then grind them.

Other Uses: Drinking a mixture of pulp and water will help cure diarrhea. Often the hollow trunks are good sources of fresh water. The bark can be cut into strips and pounded to obtain a strong fiber for making rope.

Batoko plum
Flacourtia inermis

Description: This shrub or small tree has dark green, alternate, simple leaves. Its fruits are bright red and contain six or more seeds.

Habitat and Distribution: This plant is a native of the Philippines but is widely cultivated for its fruit in other areas. It can be found in clearings and at the edges of the tropical rain forests of Africa and Asia.

Edible Parts: Eat the fruit raw or cooked.

Bearberry or kinnikinnick
Arctostaphylos uva-ursi

Description: This plant is a common evergreen shrub with reddish, scaly bark and thick, leathery leaves 4 centimeters long and 1 centimeter wide. It has white flowers and bright red fruits.

Habitat and Distribution: This plant is found in arctic, subarctic, and temperate regions, most often in sandy or rocky soil.

Edible Parts: Its berries are edible raw or cooked. You can make a refreshing tea from its young leaves.

Beech
Fagus species

Description: Beech trees are large (9 to 24 meters), symmetrical forest trees that have smooth, light gray bark and dark green foliage. The character of its bark, plus its clusters of prickly seedpods, clearly distinguish the beech tree in the field.

Habitat and Distribution: This tree is found in the Temperate Zone. It grows wild in the eastern United States, Europe, Asia, and North Africa. It is found in moist areas, mainly in the forests. This tree is common throughout southeastern Europe and across temperate Asia. Beech relatives are also found in Chile, New Guinea, and New Zealand.

Edible Parts: The mature beechnuts readily fall out of the husklike seedpods. You can eat these dark brown triangular nuts by breaking the thin shell with your fingernail and removing the white, sweet kernel inside. Beechnuts are one of the most delicious of all wild nuts. They are a most useful survival food because of the kernel's high oil content. You can also use the beechnuts as a coffee substitute. Roast them so that the kernel becomes golden brown and quite hard. Then pulverize the kernel and, after boiling or steeping in hot water, you have a passable coffee substitute.

Bignay
Antidesma bunius

Description: Bignay is a shrub or small tree, 3 to 12 meters tall, with shiny, pointed leaves about 15 centimeters long. Its flowers are small, clustered, and green. It has fleshy, dark red or black fruit and a single seed. The fruit is about 1 centimeter in diameter.

Habitat and Distribution: This plant is found in rain forests and semievergreen seasonal forests in the tropics. It is found in open places and in secondary forests. It grows wild from the Himalayas to Ceylon and eastward through Indonesia to northern Australia. However, it may be found anywhere in the tropics in cultivated forms.

Edible Parts: The fruit is edible raw. Do not eat any other parts of the tree. In Africa, the roots are toxic. Other parts of the plant may be poisonous.

CAUTION
Eaten in large quantities, the fruit may have a laxative effect.

Blackberry, raspberry, and dewberry
Rubus species

Description: These plants have prickly stems (canes) that grow upward, arching back toward the ground. They have alternate, usually compound leaves. Their fruits may be red, black, yellow, or orange.

Habitat and Distribution: These plants grow in open, sunny areas at the margin of woods, lakes, streams, and roads throughout temperate regions. There is also an arctic raspberry.

Edible Parts: The fruits and peeled young shoots are edible. Flavor varies greatly.

Other Uses: Use the leaves to make tea. To treat diarrhea, drink a tea made by brewing the dried root bark of the blackberry bush.

Blueberry and huckleberry
Vaccinium and Gaylussacia species

Description: These shrubs vary in size from 30 to 3.7 centimeters tall. All have alternate, simple leaves. Their fruits may be dark blue, black, or red and have many small seeds.

Habitat and Distribution: These plants prefer open, sunny areas. They are found throughout much of the north temperate regions and at higher elevations in Central America.

Edible Parts: Their fruits are edible raw.

Breadfruit
Artocarpus incisa

Description: This tree may grow up to 9 meters tall. It has dark green, deeply divided leaves that are 75 centimeters long and 30 centimeters wide. Its fruits are large, green, ball-like structures up to 30 centimeters across when mature.

Habitat and Distribution: Look for this tree at the margins of forests and homesites in the humid tropics. It is native to the South Pacific region but has been widely planted in the West Indies and parts of Polynesia.

Edible Parts: The fruit pulp is edible raw. The fruit can be sliced, dried, and ground into flour for later use. The seeds are edible cooked.

Other Uses: The thick sap can serve as glue and caulking material. You can also use it as birdlime (to entrap small birds by smearing the sap on twigs where they usually perch).

Burdock
Arctium lappa

Description: This plant has wavy-edged, arrow-shaped leaves and flower heads in burrlike clusters. It grows up to 2 meters tall, with purple or pink flowers and a large, fleshy root.

Habitat and Distribution: Burdock is found worldwide in the North Temperate Zone. Look for it in open waste areas during spring and summer.

Edible Parts: Peel the tender leaf stalks and eat them raw or cook them like greens. The roots are also edible boiled or baked.

CAUTION
Do not confuse burdock with rhubarb that has poisonous leaves.

Other Uses: A liquid made from the roots will help to induce sweating and increase urination. Dry the root, simmer it in water, strain the liquid, and then drink the strained liquid. Use the fiber from the dried stalk to weave cordage.

Burl Palm
Corypha elata

Description: This tree may reach 18 meters in height. It has large, fan-shaped leaves up to 3 meters long and split into about 100 narrow segments. It bears flowers in huge clusters at the top of the tree. The tree dies after flowering.

Habitat and Distribution: This tree grows in coastal areas of the East Indies.

Edible Parts: The trunk contains starch that is edible raw. The very tip of the trunk is also edible raw or cooked. You can get large quantities of liquid by bruising the flowering stalk. The kernels of the nuts are edible.

> **CAUTION**
> The seeds covering may cause dermatitis in some individuals.

Other Uses: You can use the leaves as weaving material.

Canna lily
Canna indica

Description: The canna lily is a coarse perennial herb, 90 centimeters to 3 meters tall. The plant grows from a large, thick, underground rootstock that is edible. Its large leaves resemble those of the banana plant but are not so large. The flowers of wild canna lily are usually small, relatively inconspicuous, and brightly colored reds, oranges, or yellows.

Habitat and distribution: As a wild plant, the canna lily is found in all tropical areas, especially in moist places along streams, springs, ditches, and the margins of woods. It may also be found in wet, temperate, mountainous regions. It is easy to recognize because it is commonly cultivated in flower gardens in the United States.

Edible Parts: The large and much branched rootstocks are full of edible starch. The younger parts may be finely chopped and then boiled or pulverized into a meal. Mix in the young shoots of palm cabbage for flavoring.

 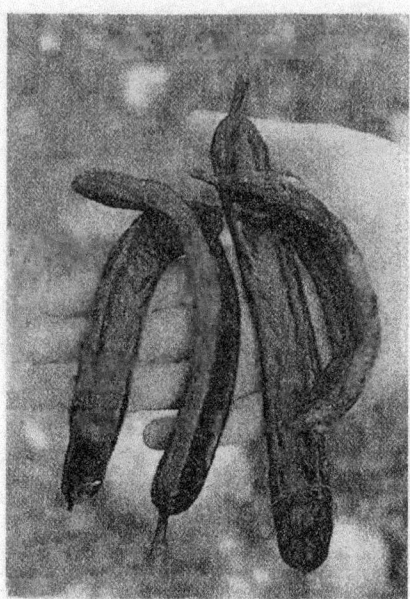

Carob tree
Ceratonia siliqua

Description: This large tree has a spreading crown. Its leaves are compound and alternate. Its seedpods, also known as Saint John's bread, are up to 45 centimeters long and are filled with round, hard seeds and a thick pulp.

Habitat and Distribution: This tree is found throughout the Mediterranean, the Middle East, and parts of North Africa.

Edible Parts: The young tender pods are edible raw or boiled. You can pulverize the seeds in mature pods and cook as porridge.

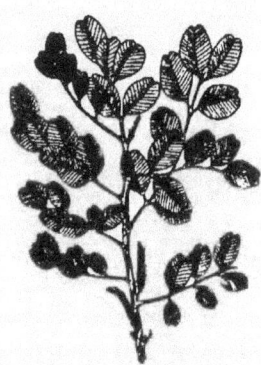

 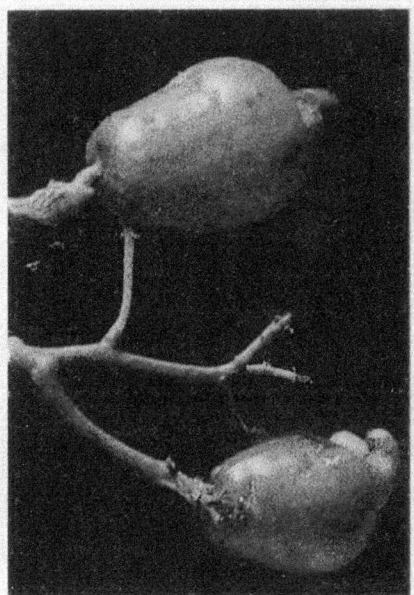

Cashew nut
Anacardium occidentale

Description: The cashew is a spreading evergreen tree growing to a height of 12 meters, with leaves up to 20 centimeters long and 10 centimeters wide. Its flowers are yellowish-pink. Its fruit is very easy to recognize because of its peculiar structure. The fruit is thick and pear-shaped, pulpy and red or yellow when ripe. This fruit bears a hard, green, kidney-shaped nut at its tip. The nut is smooth, shiny, and green or brown according to its maturity.

Habitat and Distribution: The cashew is native to the West Indies and northern South America, but transplantation has spread it to all tropical climates. In the Old World, it has escaped from cultivation and appears to be wild at least in parts of Africa and India.

Edible Parts: The nut encloses one seed. The seed is edible when roasted. The pear-shaped fruit is juicy, sweet-acid, and astringent. It is quite safe and considered delicious by most people who eat it.

CAUTION
The green hull surrounding the nut contains a resinous irritant poison that will blister the lips and tongue like poison ivy. Heat destroys this poison when roasting the nuts.

Cattail
Typha latifolia

Description: Cattails are grasslike plants with strap-shaped leaves 1 to 5 centimeters wide and growing up to 1.8 meters tall. The male flowers are borne in a dense mass above the female flowers. These last only a short time, leaving the female flowers that develop into the brown cattail. Pollen from the male flowers is often abundant and bright yellow.

Habitat and Distribution: Cattails are found throughout most of the world. Look for them in full sun areas at the margins of lakes, streams, canals, rivers, and brackish water.

Edible Parts: The young tender shoots are edible raw or cooked. The rhizome is often very tough but is a rich source of starch. Pound the rhizome to remove the starch and use as a flour. The pollen is also an exceptional source of starch. When the cattail is immature and still green, you can boil the female portion and eat it like corn on the cob.

Other Uses: The dried leaves are an excellent source of weaving material you can use to make floats and rafts. The cottony seeds make good pillow stuffing and insulation. The fluff makes excellent tinder. Dried cattails are effective insect repellents when burned.

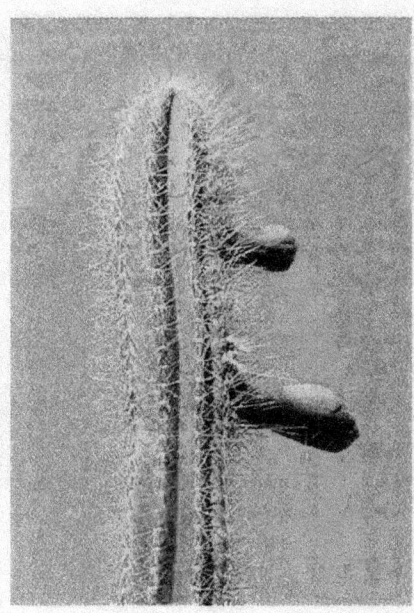

Cereus cactus
Cereus species

Description: These cacti are tall and narrow with angled stems and numerous spines.

Habitat and Distribution: They may be found in true deserts and other dry, open, sunny areas throughout the Caribbean region, Central America, and the western United States.

Edible Parts: The fruits are edible, but some may have a laxative effect.

Other Uses: The pulp of the cactus is a good source of water. Break open the stem and scoop out the pulp.

Chestnut
Castanea sativa

Description: The European chestnut is usually a large tree, up to 18 meters in height.

Habitat and Distribution: In temperate regions, the chestnut is found in both hardwood and coniferous forests. In the tropics, it is found in semievergreen seasonal forests. They are found over all of middle and south Europe and across middle Asia to China and Japan. They are relatively abundant along the edge of meadows and as a forest tree. The European chestnut is one of the most common varieties. Wild chestnuts in Asia belong to the related chestnut species.

Edible Parts: Chestnuts are highly useful as survival food. Ripe nuts are usually picked in autumn, although unripe nuts picked while green may also be used for food. Perhaps the easiest way to prepare them is to roast the ripe nuts in embers. Cooked this way, they are quite tasty, and you can eat large quantities. Another way is to boil the kernels after removing the outer shell. After being boiled until fairly soft, you can mash the nuts like potatoes.

Chicory
Cichorium intybus

Description: This plant grows up to 1.8 meters tall. It has leaves clustered at the base of the stem and some leaves on the stem. The base leaves resemble those of the dandelion. The flowers are sky blue and stay open only on sunny days. Chicory has a milky juice.

Habitat and Distribution: Look for chicory in old fields, waste areas, weedy lots, and along roads. It is a native of Europe and Asia, but is also found in Africa and most of North America where it grows as a weed.

Edible Parts: All parts are edible. Eat the young leaves as a salad or boil to eat as a vegetable. Cook the roots as a vegetable. For use as a coffee substitute, roast the roots until they are dark brown and then pulverize them.

Chufa
Cyperus esculentus

Description: This very common plant has a triangular stem and grasslike leaves. It grows to a height of 20 to 60 centimeters. The mature plant has a soft furlike bloom that extends from a whorl of leaves. Tubers 1 to 2.5 centimeters in diameter grow at the ends of the roots.

Habitat and Distribution: Chufa grows in moist sandy areas throughout the world. It is often an abundant weed in cultivated fields.

Edible Parts: The tubers are edible raw, boiled, or baked. You can also grind them and use them as a coffee substitute.

Coconut
Cocos nucifera

Description: This tree has a single, narrow, tall trunk with a cluster of very large leaves at the top. Each leaf may be over 6 meters long with over 100 pairs of leaflets.

Habitat and Distribution: Coconut palms are found throughout the tropics. They are most abundant near coastal regions.

Edible Parts: The nut is a valuable source of food. The milk of the young coconut is rich in sugar and vitamins and is an excellent source of liquid. The nut meat is also nutritious but is rich in oil. To preserve the meat, spread it in the sun until it is completely dry.

Other Uses: Use coconut oil to cook and to protect metal objects from corrosion. Also use the oil to treat saltwater sores, sunburn, and dry skin. Use the oil in improvised torches. Use the tree trunk as building material and the leaves as thatch. Hollow out the large stump for use as a food container. The coconut husks are good flotation devices and the husk's fibers are used to weave ropes and other items. Use the gauzelike fibers at the leaf bases as strainers or use them to weave a bug net or to make a pad to use on wounds. The husk makes a good abrasive. Dried husk fiber is an excellent tinder. A smoldering husk helps to repel mosquitoes. To render coconut oil, put the coconut meat in the sun, heat it over a slow fire, or boil it in a pot of water. Coconuts washed out to sea are a good source of fresh liquid for the sea survivor.

Common jujube
Ziziphus jujuba

Description: The common jujube is either a deciduous tree growing to a height of 12 meters or a large shrub, depending upon where it grows and how much water is available for growth. Its branches are usually spiny. Its reddish-brown to yellowish-green fruit is oblong to ovoid, 3 centimeters or less in diameter, smooth, and sweet in flavor, but has a rather dry pulp around a comparatively large stone. Its flowers are green.

Habitat and Distribution: The jujube is found in forested areas of temperate regions and in desert scrub and waste areas worldwide. It is common in many of the tropical and subtropical areas of the Old World. In Africa, it is found mainly bordering the Mediterranean. In Asia, it is especially common in the drier parts of India and China. The jujube is also found throughout the East Indies. It can be found bordering some desert areas.

Edible Parts: The pulp, crushed in water, makes a refreshing beverage. If time permits, you can dry the ripe fruit in the sun like dates. Its fruits are high in vitamins A and C.

Cranberry
Vaccinium macrocarpon

Description: This plant has tiny leaves arranged alternately. Its stem creeps along the ground. Its fruits are red berries.

Habitat and Distribution: It only grows in open, sunny, wet areas in the colder regions of the Northern Hemisphere.

Edible Parts: The berries are very tart when eaten raw. Cook in a small amount of water and add sugar, if available, to make a jelly.

Other Uses: Cranberries may act as a diuretic. They are useful for treating urinary tract infections.

Crowberry
Empetrum nigrum

Description: This is a dwarf evergreen shrub with short needlelike leaves. It has small, shiny, black berries that remain on the bush through the winter.

Habitat and Distribution: Look for this plant in tundra throughout arctic regions of North America and Eurasia.

Edible Parts: The fruits are edible fresh or can be dried for later use.

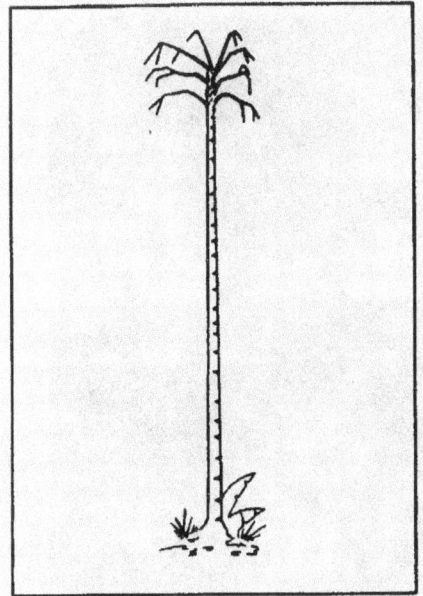 

Cuipo tree
Cavanillesia platanifolia

Description: This is a very dominant and easily detected tree because it extends above the other trees. Its height ranges from 45 to 60 meters. It has leaves only at the top and is bare 11 months out of the year. It has rings on its bark that extend to the top to make it easily recognizable. Its bark is reddish or gray in color. Its roots are light reddish-brown or yellowish-brown.

Habitat and Distribution: The cuipo tree is located primarily in Central American tropical rain forests in mountainous areas.

Edible Parts: To get water from this tree, cut a piece of the root and clean the dirt and bark off one end, keeping the root horizontal. Put the clean end to your mouth or canteen and raise the other. The water from this tree tastes like potato water.

Other Uses: Use young saplings and the branches' inner bark to make rope.

Dandelion
Taraxacum officinale

Description: Dandelion leaves have a jagged edge, grow close to the ground, and are seldom more than 20 centimeters long. Its flowers are bright yellow. There are several dandelion species.

Habitat and Distribution: Dandelions grow in open, sunny locations throughout the Northern Hemisphere.

Edible Parts: All parts are edible. Eat the leaves raw or cooked. Boil the roots as a vegetable. Roots roasted and ground are a good coffee substitute. Dandelions are high in vitamins A and C and in calcium.

Other Uses: Use the white juice in the flower stems as glue.

Date Palm
Phoenix dactylifera

Description: The date palm is a tall, unbranched tree with a crown of huge, compound leaves. Its fruit is yellow when ripe.

Habitat and Distribution: This tree grows in arid semitropical regions. It is native to North Africa and the Middle East but has been planted in the arid semitropics in other parts of the world.

Edible Parts: The fruit is edible fresh but is very bitter if eaten before it is ripe. You can dry the fruits in the sun and preserve them for a long time.

Other Uses: The trunks provide valuable building material in deserts regions where few other treelike plants are found. The leaves are durable and you can use them for thatching and as weaving material. The base of the leaves resembles coarse cloth that you can use for scrubbing and cleaning.

Daylily
Hemerocallis fulva

Description: This plant has unspotted, tawny blossoms that open for 1 day only. It has long, swordlike, green basal leaves. Its root is a mass of swollen and elongated tubers.

Habitat and Distribution: Daylilies are found worldwide in Tropic and Temperate Zones. They are grown as a vegetable in the Orient and as an ornamental plant elsewhere.

Edible Parts: The young green leaves are edible raw or cooked. Tubers are also edible raw or cooked. You can eat its flowers raw, but they taste better cooked. You can also fry the flowers for storage.

CAUTION
Eating excessive amounts of raw flowers may cause diarrhea.

Duchesnea or Indian strawberry
Duchesnea indica

Description: The duchesnea is a small plant that has runners and three-parted leaves. Its flowers are yellow and its fruit resembles a strawberry.

Habitat and distribution: It is native to southern Asia but is a common weed in warmer temperate regions. Look for it in lawns, gardens, and along roads.

Edible Parts: Its fruit is edible. Eat it fresh.

Elderberry
Sambucus canadensis

Description: Elderberry is a many-stemmed shrub with opposite, compound leaves. It grows to a height of 6 meters. Its flowers are fragrant, white, and borne in large flat-topped clusters up to 30 centimeters across. Its berrylike fruits are dark blue or black when ripe.

Habitat and Distribution: This plant is found in open, usually wet areas at the margins of marshes, rivers, ditches, and lakes. It grows throughout much of eastern North America and Canada.

Edible Parts: The flowers and fruits are edible. You can make a drink by soaking the flower heads for 8 hours, discarding the flowers, and drinking the liquid.

CAUTION
All other parts of the plant are poisonous and dangerous if eaten.

Fireweed
Epilobium angustifolium

Description: This plant grows up to 1.8 meters tall. It has large, showy, pink flowers and lance-shaped leaves. Its relative, the dwarf fireweed (*Epilobium latifolium*), grows 30 to 60 centimeters tall.

Habitat and Distribution: Tall fireweed is found in open woods, on hillsides, on stream banks, and near seashores in arctic region. It is especially abundant in burned-over areas. Dwarf fireweed is found along streams, sandbars, and lakeshores and on alpine and arctic slopes.

Edible Parts: The leaves, stems, and flowers are edible in the spring but become tough in summer. You can split open the stems of old plants and eat the pith raw.

Fishtail palm
Caryota urens

Description: Fishtail palms are large trees, at least 18 meters tall. Their leaves are unlike those of any other palm; the leaflets are irregular and toothed on the upper margins. All other palms have either fan-shaped or featherlike leaves. Its massive flowering shoot is borne at the top of the tree and hangs downward.

Habitat and Distribution: The fishtail palm is native to the tropics of India, Assam, and Burma. Several related species also exist in Southeast Asia and the Philippines. These palms are found in open hill country and jungle areas.

Edible Parts: The chief food in this palm is the starch stored in large quantities in its trunk. The juice from the fishtail palm is very nourishing and you have to drink it shortly after getting it from the palm flower shoot. Boil the juice down to get a rich sugar syrup. Use the same method as for the sugar palm to get the juice. The palm cabbage may be eaten raw or cooked.

Foxtail grass
Setaria species

Description: This weedy grass is readily recognizable by the narrow, cylindrical head containing long hairs. Its grains are small, less than 6 millimeters long. The dense heads of grain often droop when ripe.

Habitat and Distribution: Look for foxtail grasses in open, sunny areas, along roads, and at the margins of fields. Some species occur in wet, marshy areas. Species of *Setaria* are found throughout the United States, Europe, western Asia, and tropical Africa. In some parts of the world, foxtail grasses are grown as a food crop.

Edible Parts: The grains are edible raw but are very hard and sometimes bitter. Boiling removes some of the bitterness and makes them easier to eat.

Goa Bean
Psophocarpus tetragonolobus

Description: The goa bean is a climbing plant that may cover small shrubs and trees. Its bean pods are 22 centimeters long, its leaves 15 centimeters long, and its flowers are bright blue. The mature pods are 4-angled, with jagged wings on the pods.

Habitat and Distribution: This plant grows in tropical Africa, Asia, the East Indies, the Philippines, and Taiwan. This member of the bean (legume) family serves to illustrate a kind of edible bean common in the tropics of the Old World. Wild edible beans of this sort are most frequently found in clearings and around abandoned garden sites. They are more rare in forested areas.

Edible Parts: You can eat the young pods like string beans. The mature seeds are a valuable source of protein after parching or roasting them over hot coals. You can germinate the seeds (as you can many kinds of beans) in damp moss and eat the resultant sprouts. The thickened roots are edible raw. They are slightly sweet, with the firmness of an apple. You can also eat the young leaves as a vegetable, raw or steamed.

Hackberry
Celtis species

Description: Hackberry trees have smooth, gray bark that often has corky warts or ridges. The tree may reach 39 meters in height. Hackberry trees have long-pointed leaves that grow in two rows. This tree bears small, round berries that can be eaten when they are ripe and fall from the tree. The wood of the hackberry is yellowish.

Habitat and Description: This plant is widespread in the United States, especially in and near ponds.

Edible Parts: Its berries are edible when they are ripe and fall from the tree.

Hazelnut or wild filbert
Corylus species

Description: Hazelnuts grow on bushes 1.8 to 3.6 meters high. One species in Turkey and another in China are large trees. The nut itself grows in a very bristly husk that conspicuously contracts above the nut into a long neck. The different species vary in this respect as to size and shape.

Habitat and Distribution: Hazelnuts are found over wide areas in the United States, especially the eastern half of the country and along the Pacific coast. These nuts are also found in Europe where they are known as filberts. The hazelnut is common in Asia, especially in eastern Asia from the Himalayas to China and Japan. The hazelnut usually grows in the dense thickets along stream banks and open places. They are not plants of the dense forest.

Edible Parts: Hazelnuts ripen in the autumn when you can crack them open and eat the kernel. The dried nut is extremely delicious. The nut's high oil content makes it a good survival food. In the unripe stage, you can crack them open and eat the fresh kernel.

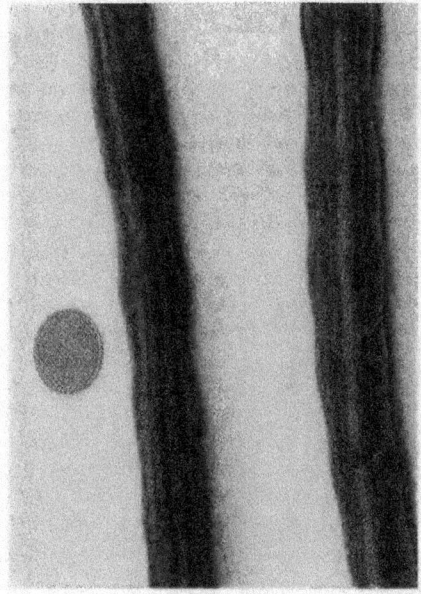
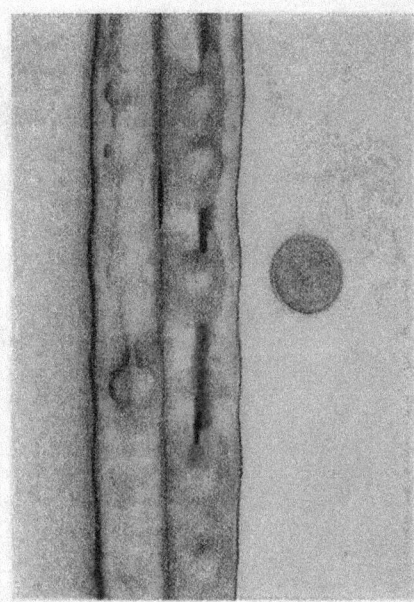

Horseradish tree
Moringa pterygosperma

Description: This tree grows from 4.5 to 14 meters tall. Its leaves have a fernlike appearance. Its flowers and long, pendulous fruits grow on the ends of the branches. Its fruit (pod) looks like a giant bean. Its 25-to-60-centimeter-long-pods are triangular in cross section, with strong ribs. Its roots have a pungent odor.

Habitat and Distribution: This tree is found in the rain forests and semievergreen seasonal forests of the tropical regions. It is widespread in India, Southeast Asia, Africa, and Central America. Look for it in abandoned fields and gardens and at the edges of forests.

Edible Parts: The leaves are edible raw or cooked, depending on their hardness. Cut the young seedpods into short lengths and cook them like string beans or fry them. You can get oil for frying by boiling the young fruits of palms and skimming the oil off the surface of the water. You can eat the flowers as part of a salad. You can chew fresh, young seedpods to eat the pulpy and soft seeds. The roots may be ground as a substitute for seasoning similar to horseradish.

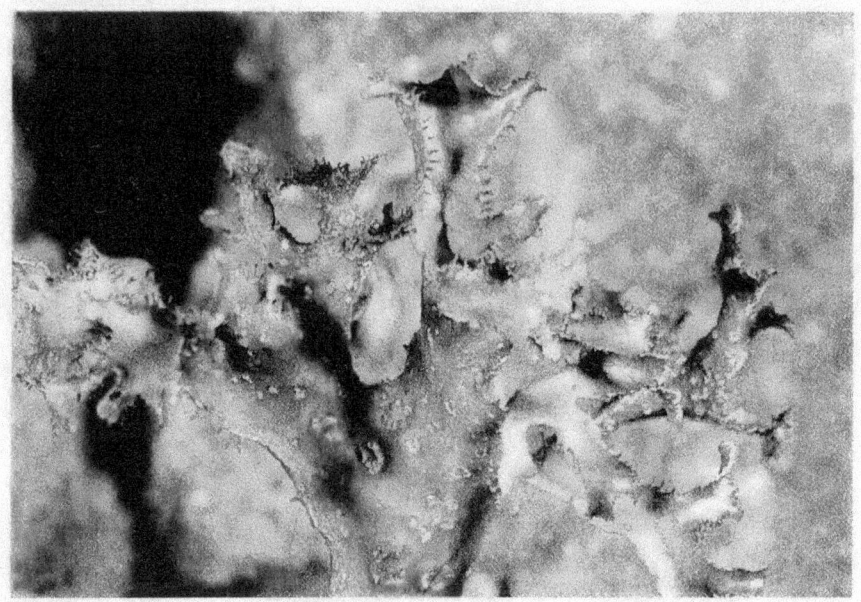

Iceland moss
Cetraria islandica

Description: This moss grows only a few inches high. Its color may be gray, white, or even reddish.

Habitat and Distribution: Look for it in open areas. It is found only in the arctic.

Edible Parts: All parts of the Iceland moss are edible. During the winter or dry season, it is dry and crunchy but softens when soaked. Boil the moss to remove the bitterness. After boiling, eat by itself, or add to milk or grains as a thickening agent. Dried plants store well.

Indian potato or Eskimo potato
Claytonia species

Description: All *Claytonia* species are somewhat fleshy plants only a few centimeters tall, with showy flowers about 2.5 centimeters across.

Habitat and Distribution: Some species are found in rich forests where they are conspicuous before the leaves develop. Western species are found throughout most of the northern United States and in Canada.

Edible Parts: The tubers are edible but you should boil them before eating.

Juniper
Juniperus species

Description: Junipers, sometimes called cedars, are trees or shrubs with very small, scalelike leaves densely crowded around the branches. Each leaf is less than 1.2 centimeters long. All species have a distinct aroma resembling the well-known cedar. The berrylike cones are usually blue and covered with whitish wax.

Habitat and Distribution: Look for junipers in open, dry, sunny areas throughout North America and northern Europe. Some species are found in southeastern Europe, across Asia to Japan, and in the mountains of North Africa.

Edible Parts: The berries and twigs are edible. Eat the berries raw or roast the seeds to use as a coffee substitute. Use dried and crushed berries as a seasoning for meat. Gather young twigs to make a tea.

CAUTION
Many plants may be called cedars but are not related to junipers and may be harmful. Always look for the berrylike structures, needle leaves, and resinous, fragrant sap to be sure the plant you have is a juniper.

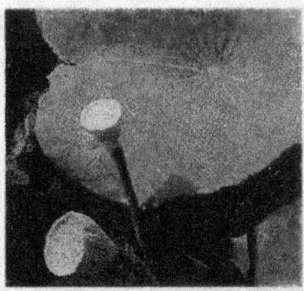

Lotus
Nelumbo species

Description: There are two species of lotus: one has yellow flowers and the other pink flowers. The flowers are large and showy. The leaves, which may float on or rise above the surface of the water, often reach 1.5 meters in radius. The fruit has a distinctive flattened shape and contains up to 20 hard seeds.

Habitat and Distribution: The yellow-flowered lotus is native to North America. The pink-flowered species, which is widespread in the Orient, is planted in many other areas of the world. Lotuses are found in quiet fresh water.

Edible Parts: All parts of the plant are edible raw or cooked. The underwater parts contain large quantities of starch. Dig the fleshy portions from the mud and bake or boil them. Boil the young leaves and eat them as a vegetable. The seeds have a pleasant flavor and are nutritious. Eat them raw, or parch and grind them into flour.

Malanga
Xanthosoma caracu

Description: This plant has soft, arrow-shaped leaves, up to 60 centimeters long. The leaves have no aboveground stems.

Habitat and Distribution: This plant grows widely in the Caribbean region. Look for it in open, sunny fields.

Edible Parts: The tubers are rich in starch. Cook them before eating to destroy a poison contained in all parts of the plant.

> **WARNING**
> Always cook before eating.

Mango
Mangifera indica

Description: This tree may reach 30 meters in height. It has alternate, simple, shiny, dark green leaves. Its flowers are small and inconspicuous. Its fruits have a large single seed. There are many cultivated varieties of mango. Some have red flesh, others yellow or orange, often with many fibers and a kerosene taste.

Habitat and Distribution: This tree grows in warm, moist regions. It is native to northern India, Burma, and western Malaysia. It is now grown throughout the tropics.

Edible Parts: The fruits are a nutritious food source. The unripe fruit can be peeled and its flesh eaten by shredding it and eating it like a salad. The ripe fruit can be peeled and eaten raw. Roasted seed kernels are edible.

CAUTION
If you are sensitive to poison ivy, avoid eating mangoes, as they cause a severe reaction in sensitive individuals.

Manioc
Manihot utillissima

Description: Manioc is a perennial shrubby plant, 1 to 3 meters tall, with jointed stems and deep green, fingerlike leaves. It has large, fleshy rootstocks.

Habitat and Distribution: Manioc is widespread in all tropical climates, particularly in moist areas. Although cultivated extensively, it may be found in abandoned gardens and growing wild in many areas.

Edible Parts: The rootstocks are full of starch and high in food value. Two kinds of manioc are known: bitter and sweet. Both are edible. The bitter type contains poisonous hydrocyanic acid. To prepare manioc, first grind the fresh manioc roots into a pulp, then cook it for at least 1 hour to remove the bitter poison from the roots. Then flatten the pulp into cakes and bake as bread. Manioc cakes or flour will keep almost indefinitely if protected against insects and dampness. Wrap them in banana leaves for protection.

CAUTION
For safety, always cook the roots of either type.

Marsh marigold
Caltha palustris

Description: This plant has rounded, dark green leaves arising from a short stem. It has bright yellow flowers.

Habitat and Distribution: This plant is found in bogs, lakes, and slow-moving streams. It is abundant in arctic and subarctic regions and in much of the eastern regions of the northern United States.

Edible Parts: All parts are edible if boiled.

CAUTION
As with all water plants, do not eat this plant raw. Raw water plants may carry dangerous organisms that are removed only by cooking.

Mulberry
Morus species

Description: This tree has alternate, simple, often lobed leaves with rough surfaces. Its fruits are blue or black and many seeded.

Habitat and Distribution: Mulberry trees are found in forests, along roadsides, and in abandoned fields in Temperate and Tropical Zones of North America, South America, Europe, Asia, and Africa.

Edible Parts: The fruit is edible raw or cooked. It can be dried for eating later.

> **CAUTION**
> When eaten in quantity, mulberry fruit acts as a laxative. Green, unripe fruit can be hallucinogenic and cause extreme nausea and cramps.

Other Uses: You can shred the inner bark of the tree and use it to make twine or cord.

Nettle
Urtica and *Laportea* species

Description: These plants grow several feet high. They have small, inconspicuous flowers. Fine, hairlike bristles cover the stems, leafstalks, and undersides of leaves. The bristles cause a stinging sensation when they touch the skin.

Habitat and Distribution: Nettles prefer moist areas along streams or at the margins of forests. They are found throughout North America, Central America, the Caribbean, and northern Europe.

Edible Parts: Young shoots and leaves are edible. Boiling the plant for 10 to 15 minutes destroys the stinging element of the bristles. This plant is very nutritious.

Other Uses: Mature stems have a fibrous layer that you can divide into individual fibers and use to weave string or twine.

Nipa palm
Nipa fruticans

Description: This palm has a short, mainly underground trunk and very large, erect leaves up to 6 meters tall. The leaves are divided into leaflets. A flowering head forms on a short erect stem that rises among the palm leaves. The fruiting (seed) head is dark brown and may be 30 centimeters in diameter.

Habitat and Distribution: This palm is common on muddy shores in coastal regions throughout eastern Asia.

Edible Parts: The young flower stalk and the seeds provide a good source of water and food. Cut the flower stalk and collect the juice. The juice is rich in sugar. The seeds are hard but edible.

Other Uses: The leaves are excellent as thatch and coarse weaving material.

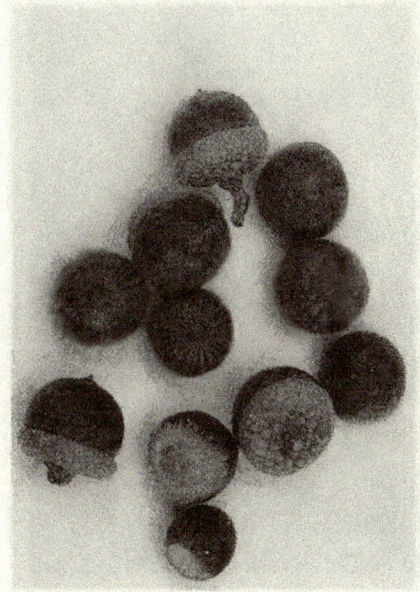

Oak
Quercus species

Description: Oak trees have alternate leaves and acorn fruits. There are two main groups of oaks: red and white. The red oak group has leaves with bristles and smooth bark in the upper part of the tree. Red oak acorns take 2 years to mature. The white oak group has leaves without bristles and a rough bark in the upper portion of the tree. White oak acorns mature in 1 year.

Habitat and Distribution: Oak trees are found in many habitats throughout North America, Central America, and parts of Europe and Asia.

Edible Parts: All parts are edible, but often contain large quantities of bitter substances. White oak acorns usually have a better flavor than red oak acorns. Gather and shell the acorns. Soak red oak acorns in water for 1 to 2 days to remove the bitter substance. You can speed up this process by putting wood ashes in the water in which you soak the acorns. Boil the acorns or grind them into flour and use the flour for baking. You can use acorns that you baked until very dark as a coffee substitute.

CAUTION
Tannic acid gives the acorns their bitter taste. Eating an excessive amount of acorns high in tannic acid can lead to kidney failure. Before eating acorns, leach out this chemical.

Orach
Atriplex species

Description: This plant is vinelike in growth and has arrowhead-shaped, alternate leaves up to 5 centimeters long. Young leaves may be silver-colored. Its flowers and fruits are small and inconspicuous.

Habitat and Distribution: Orach species are entirely restricted to salty soils. They are found along North America's coasts and on the shores of alkaline lakes inland. They are also found along seashores from the Mediterranean countries to inland areas in North Africa and eastward to Turkey and central Siberia.

Edible Parts: The entire plant is edible raw or boiled.

Palmetto palm
Sabal palmetto

Description: The palmetto palm is a tall, unbranched tree with persistent leaf bases on most of the trunk. The leaves are large, simple, and palmately lobed. Its fruits are dark blue or black with a hard seed.

Habitat and Distribution: The palmetto palm is found throughout the coastal regions of the southeastern United States.

Edible Parts: The fruits are edible raw. The hard seeds may be ground into flour. The heart of the palm is a nutritious food source at any time. Cut off the top of the tree to obtain the palm heart.

Papaya or pawpaw
Carica papaya

Description: The papaya is a small tree 1.8 to 6 meters tall, with a soft, hollow trunk. When cut, the entire plant exudes a milky juice. The trunk is rough, and the leaves are crowded at the trunk's apex. The fruit grows directly from the trunk, among and below the leaves. The fruit is green before ripening. When ripe, it turns yellow or remains greenish with a squashlike appearance.

Habitat and Distribution: Papaya is found in rain forests and semievergreen seasonal forests in tropical regions and in some temperate regions as well. Look for it in moist areas near clearings and former habitations. It is also found in open, sunny places in uninhabited jungle areas.

Edible Parts: The ripe fruit is high in Vitamin C. Eat it raw or cook it like squash. Place green fruit in the sun to make it ripen quickly. Cook the young papaya leaves, flowers, and stems carefully, changing the water as for taro.

CAUTION
Be careful not to get the milky sap from the unripe fruit into your eyes. It will cause intense pain and temporary—sometimes even permanent—blindness.

Other Uses: Use the milky juice of the unripe fruit to tenderize tough meat. Rub the juice on the meat.

Persimmon
Diospyros virginiana and other species

Description: These trees have alternate, dark green, elliptic leaves with entire margins. The flowers are inconspicuous. The fruit are orange, have a sticky consistency, and have several seeds.

Habitat and Distribution: The persimmon is a common forest margin tree. It is widespread in Africa, eastern North America, and the Far East.

Edible Parts: The leaves are a good source of vitamin C. The fruits are edible raw or baked. To make tea, dry the leaves and soak them in hot water. You can eat the roasted seeds.

CAUTION
Some persons are unable to digest persimmon pulp. Unripe persimmons are highly astringent and inedible.

Pincushion cactus
Mammillaria species

Description: Members of this cactus group are round, short, barrel-shaped, and without leaves. Sharp spines cover the entire plant.

Habitat and Distribution: These cacti are found throughout much of the desert regions of the western United States and parts of Central America.

Edible Parts: They are a good source of water in the desert.

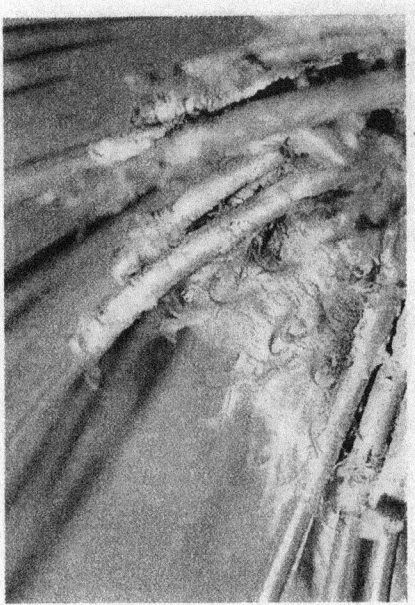

Pine
Pinus species

Description: Pine trees are easily recognized by their needlelike leaves grouped in bundles. Each bundle may contain one to five needles, the number varying among species. The tree's odor and sticky sap provide a simple way to distinguish pines from similar looking trees with needlelike leaves.

Habitat and Distribution: Pines prefer open, sunny areas. They are found throughout North America, Central America, much of the Caribbean region, North Africa, the Middle East, Europe, and some places in Asia.

Edible Parts: The seeds of all species are edible. You can collect the young male cones, which grow only in the spring, as a survival food. Boil or bake the young cones. The bark of young twigs is edible. Peel off the bark of thin twigs. You can chew the juicy inner bark; it is rich in sugar and vitamins. Eat the seeds raw or cooked. Green pine needle tea is high in vitamin C.

Other Uses: Use the resin to waterproof articles. Also use it as glue. Collect the resin from the tree. If there is not enough resin on the tree, cut a notch in the bark so more sap will seep out. Put the resin in a container and heat it. The hot resin is your glue. Use it as is or add a small amount of ash dust to strengthen it. Use it immediately. You can use hardened pine resin as an emergency dental filling.

Plantain, broad and narrow leaf
Plantago species

Description: The broad leaf plantain has leaves over 2.5 centimeters across that grow close to the ground. The flowers are on a spike that rises from the middle of the cluster of leaves. The narrow leaf plantain has leaves up to 12 centimeters long and 2.5 centimeters wide, covered with hairs. The leaves form a rosette. The flowers are small and inconspicuous.

Habitat and Distribution: Look for these plants in lawns and along roads in the North Temperate Zone. This plant is a common weed throughout much of the world.

Edible Parts: The young tender leaves are edible raw. Older leaves should be cooked. Seeds are edible raw or roasted.

Other Uses: To relieve pain from wounds or sores, wash and soak the entire plant for a short time and apply it to the injured area. To treat diarrhea, drink tea made from 28 grams (1 ounce) of the plant leaves boiled in 0.5 liter of water. The seeds and seed husks act as laxatives.

Pokeweed
Phytolacca americana

Description: This plant may grow as high as 3 meters. Its leaves are elliptic and up to 1 meter in length. It produces many large clusters of purple fruits in late spring.

Habitat and Distribution: Look for this plant in open, sunny areas in forest clearings, in fields, and along roadsides in eastern North America, Central America, and the Caribbean.

Edible Parts: The young leaves and stems are edible cooked. Boil them twice, discarding the water from the first boiling. The fruits are edible if cooked.

> **CAUTION**
> All parts of this plant are poisonous if eaten raw. Never eat the underground portions of the plant as these contain the highest concentrations of the poisons. Do not eat any plant over 25 centimeters tall or when red is showing in the plant.

Other Uses: Use the juice of fresh berries as a dye.

 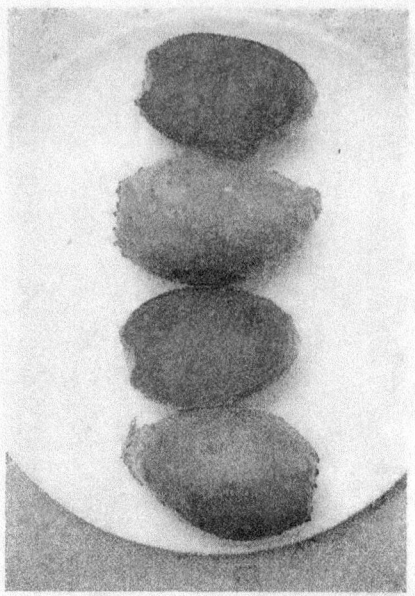

Prickly pear cactus
Opuntia species

Description: This cactus has flat, padlike stems that are green. Many round, furry dots that contain sharp-pointed hairs cover these stems.

Habitat and Distribution: This cactus is found in arid and semiarid regions and in dry, sandy areas of wetter regions throughout most of the United States and Central and South America. Some species are planted in arid and semiarid regions of other parts of the world.

Edible Parts: All parts of the plant are edible. Peel the fruits and eat them fresh or crush them to prepare a refreshing drink. Avoid the tiny, pointed hairs. Roast the seeds and grind them to a flour.

CAUTION
Avoid any prickly pear cactuslike plant with milky sap.

Other Uses: The pad is a good source of water. Peel it carefully to remove all sharp hairs before putting it in your mouth. You can also use the pads to promote healing. Split them and apply the pulp to wounds.

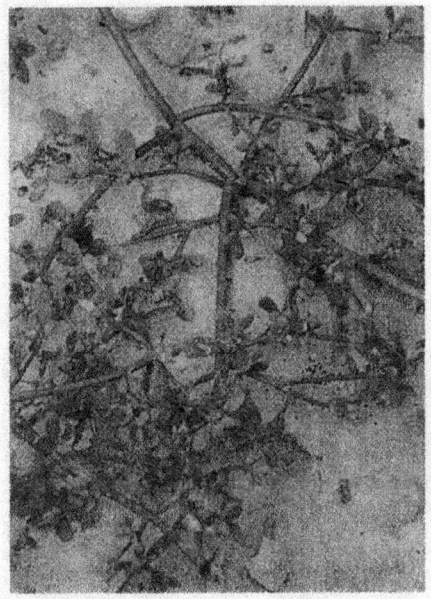 

Purslane
Portulaca oleracea

Description: This plant grows close to the ground. It is seldom more than a few centimeters tall. Its stems and leaves are fleshy and often tinged with red. It has paddleshaped leaves, 2.5 centimeters or less long, clustered at the tips of the stems. Its flowers are yellow or pink. Its seeds are tiny and black.

Habitat and Distribution: It grows in full sun in cultivated fields, field margins, and other weedy areas throughout the world.

Edible Parts: All parts are edible. Wash and boil the plants for a tasty vegetable or eat them raw. Use the seeds as a flour substitute or eat them raw.

Rattan palm
Calamus species

Description: The rattan palm is a stout, robust climber. It has hooks on the midrib of its leaves that it uses to remain attached to trees on which it grows. Sometimes, mature stems grow to 90 meters. It has alternate, compound leaves and a whitish flower.

Habitat and Distribution: The rattan palm is found from tropical Africa through Asia to the East Indies and Australia. It grows mainly in rain forests.

Edible Parts: Rattan palms hold a considerable amount of starch in their young stem tips. You can eat them roasted or raw. In other kinds, a gelatinous pulp, either sweet or sour, surrounds the seeds. You can suck out this pulp. The palm heart is also edible raw or cooked.

Other Uses: You can obtain large amounts of potable water by cutting the ends of the long stems. The stems can be used to make baskets and fish traps.

Reed
Phragmites australis

Description: This tall, coarse grass grows to 3.5 meters tall and has gray-green leaves about 4 centimeters wide. It has large masses of brown flower branches in early summer. These rarely produce grain and become fluffy, gray masses late in the season.

Habitat and Distribution: Look for reed in any open, wet area, especially one that has been disturbed through dredging. Reed is found throughout the temperate regions of both the Northern and Southern Hemispheres.

Edible Parts: All parts of the plant are edible raw or cooked in any season. Harvest the stems as they emerge from the soil and boil them. You can also harvest them just before they produce flowers, then dry and beat them into flour. You can also dig up and boil the underground stems, but they are often tough. Seeds are edible raw or boiled, but they are rarely found.

Reindeer moss
Cladonia rangiferin

Description: Reindeer moss is a low-growing plant only a few centimeters tall. It does not flower but does produce bright red reproductive structures.

Habitat and Distribution: Look for this lichen in open, dry areas. It is very common in much of North America.

Edible Parts: The entire plant is edible but has a crunchy, brittle texture. Soak the plant in water with some wood ashes to remove the bitterness, then dry, crush, and add it to milk or to other food.

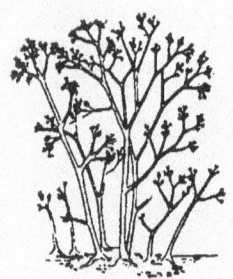

 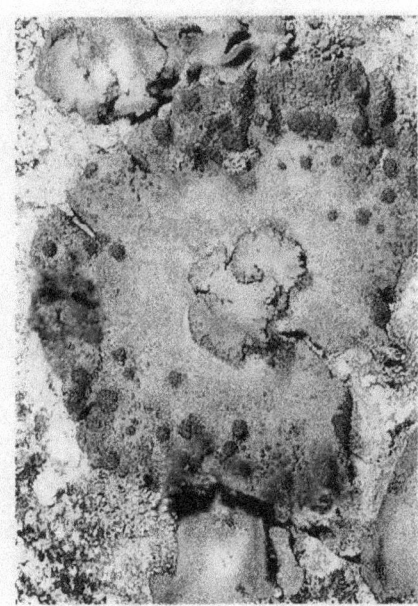

Rock tripe
Umbilicaria species

Description: This plant forms large patches with curling edges. The top of the plant is usually black. The underside is lighter in color.

Habitat and Distribution: Look on rocks and boulders for this plant. It is common throughout North America.

Edible Parts: The entire plant is edible. Scrape it off the rock and wash it to remove grit. The plant may be dry and crunchy; soak it in water until it becomes soft. Rock tripes may contain large quantities of bitter substances; soaking or boiling them in several changes of water will remove the bitterness.

CAUTION
There are some reports of poisoning from rock tripe, so apply the Universal Edibility Test.

Rose apple
Eugenia jambos

Description: This tree grows 3 to 9 meters high. It has opposite, simple, dark green, shiny leaves. When fresh, it has fluffy, yellowish-green flowers and red to purple egg-shaped fruit.

Habitat and Distribution: This tree is widely planted in all of the tropics. It can also be found in a semiwild state in thickets, waste places, and secondary forests.

Edible Parts: The entire fruit is edible raw or cooked.

Sago palm
Metroxylon sagu

Description: These palms are low trees, rarely over 9 meters tall, with a stout, spiny trunk. The outer rind is about 5 centimeters thick and hard as bamboo. The rind encloses a spongy inner pith containing a high proportion of starch. It has typical palmlike leaves clustered at the tip.

Habitat and Distribution: Sago palm is found in tropical rain forests. It flourishes in damp lowlands in the Malay Peninsula, New Guinea, Indonesia, the Philippines, and adjacent islands. It is found mainly in swamps and along streams, lakes, and rivers.

Edible Parts: These palms, when available, are of great use to the survivor. One trunk, cut just before it flowers, will yield enough sago to feed a person for 1 year. Obtain sago starch from nonflowering palms. To extract the edible sago, cut away the bark lengthwise from one half of the trunk, and pound the soft, whitish inner part (pith) as fine as possible. Knead the pith in water and strain it through a coarse cloth into a container. The fine, white sago will settle in the container. Once the sago settles, it is ready for use. Squeeze off the excess water and let it dry. Cook it as pancakes or oatmeal. Two kilograms of sago is the nutritional equivalent of 1.5 kilograms of rice. The upper part of the trunk's core does not yield sago, but you can roast it in lumps over a fire. You can also eat the young sago nuts and the growing shoots or palm cabbage.

Other Uses: Use the stems of tall sorghums as thatching materials.

 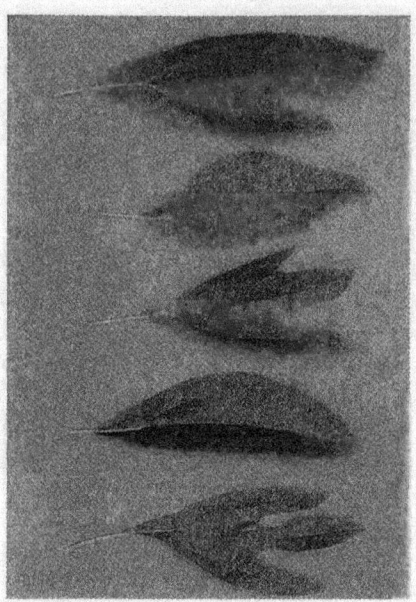

Sassafras
Sassafras albidum

Description: This shrub or small tree bears different leaves on the same plant. Some leaves will have one lobe, some two lobes, and some no lobes. The flowers, which appear in early spring, are small and yellow. The fruits are dark blue. The plant parts have a characteristic root beer smell.

Habitat and Distribution: Sassafras grows at the margins of roads and forests, usually in open, sunny areas. It is a common tree throughout eastern North America.

Edible Parts: The young twigs and leaves are edible fresh or dried. You can add dried young twigs and leaves to soups. Dig the underground portion, peel off the bark, and let it dry. Then boil it in water to prepare sassafras tea.

Other Uses: Shred the tender twigs for use as a toothbrush.

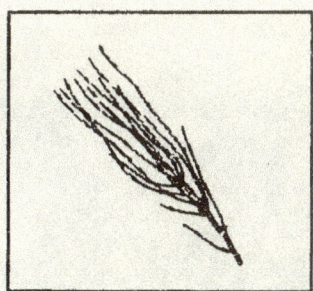
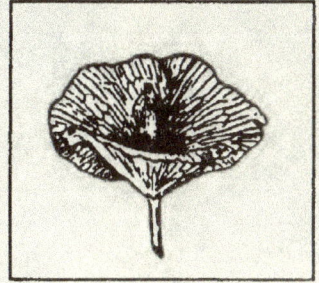

Saxual
Haloxylon ammondendron

Description: The saxual is found either as a small tree or as a large shrub with heavy, coarse wood and spongy, water-soaked bark. The branches of the young trees are vivid green and pendulous. The flowers are small and yellow.

Habitat and Distribution: The saxual is found in desert and arid areas. It is found on the arid salt deserts of Central Asia, particularly in the Turkestan region and east of the Caspian Sea.

Edible Parts: The thick bark acts as a water storage organ. You can get drinking water by pressing quantities of the bark. This plant is an important source of water in the arid regions in which it grows.

Screw pine
Pandanus species

Description: The screw pine is a strange plant built on stilts, or prop roots, that support the plant aboveground so that it appears more or less suspended in midair. These plants are either shrubby or treelike, 3 to 9 meters tall, with stiff leaves having sawlike edges. The fruits are large, roughened balls resembling pineapples, but without the tuft of leaves at the end.

Habitat and Distribution: The screw pine is a tropical plant that grows in rain forests and semievergreen seasonal forests. It is found mainly along seashores, although certain kinds occur inland for some distance, from Madagascar to southern Asia and the islands of the southwestern Pacific. There are about 180 types.

Edible Parts: Knock the ripe fruit to the ground to separate the fruit segments from the hard outer covering. Chew the inner fleshy part. Cook fruit that is not fully ripe in an earth oven. Before cooking, wrap the whole fruit in banana leaves, breadfruit leaves, or any other suitable thick, leathery leaves. After cooking for about 2 hours, you can chew fruit segments like ripe fruit. Green fruit is inedible.

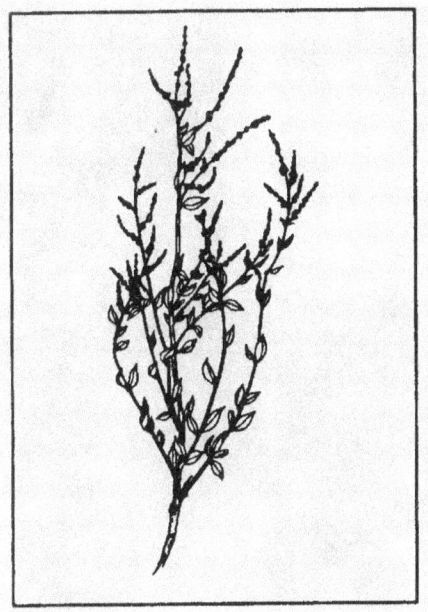 

Sea orach
Atriplex halimus

Description: The sea orach is a sparingly branched herbaceous plant with small, gray-colored leaves up to 2.5 centimeters long. Sea orach resembles lamb's-quarter, a common weed in most gardens in the United States. It produces its flowers in narrow, densely compacted spikes at the tips of its branches.

Habitat and Distribution: The sea orach is found in highly alkaline and salty areas along seashores from the Mediterranean countries to inland areas in North Africa and eastward to Turkey and central Siberia. Generally, it can be found in tropical scrub and thorn forests, steppes in temperate regions, and most desert scrub and waste areas.

Edible Parts: Its leaves are edible. In the areas where it grows, it has the healthy reputation of being one of the few native plants that can sustain man in times of want.

 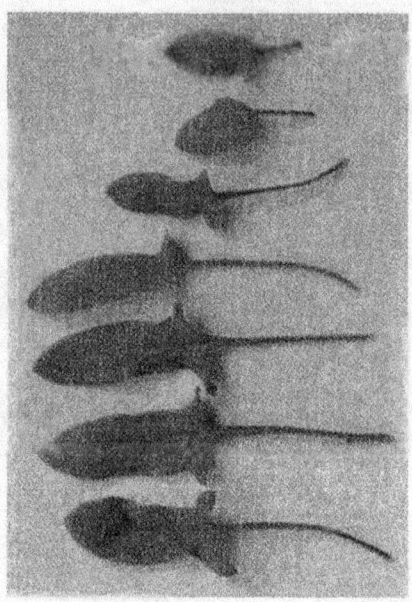

Sheep sorrel
Rumex acetosella

Description: These plants are seldom more than 30 centimeters tall. They have alternate leaves, often with arrowlike bases, very small flowers, and frequently reddish sterns.

Habitat and Distribution: Look for these plants in old fields and other disturbed areas in North America and Europe.

Edible Parts: The plants are edible raw or cooked.

> **CAUTION:**
> These plants contain oxalic acid that can be damaging if too many plants are eaten raw. Cooking seems to destroy the chemical.

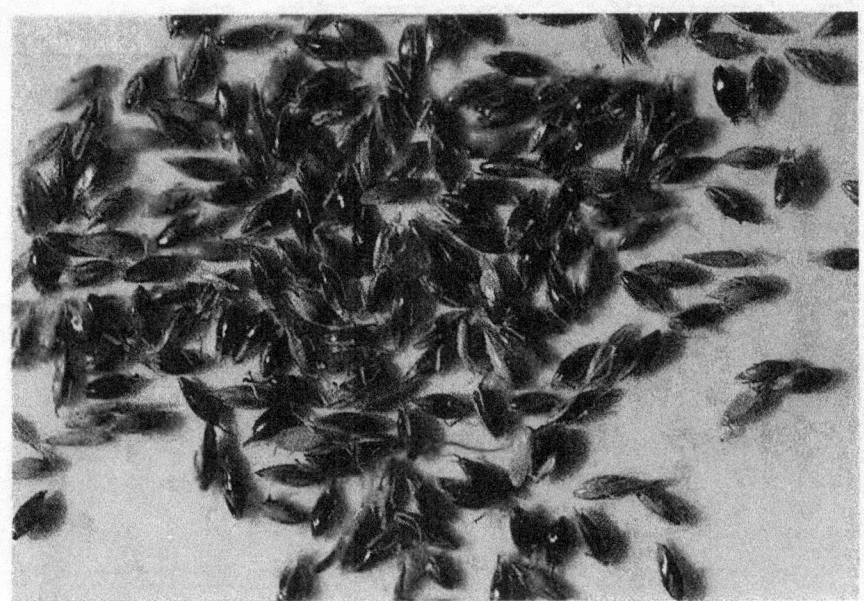

Sorghum
Sorghum species

Description: There are many different kinds of sorghum, all of which bear grains in heads at the top of the plants. The grains are brown, white, red, or black. Sorghum is the main food crop in many parts of the world.

Habitat and Distribution: Sorghum is found worldwide, usually in warmer climates. All species are found in open, sunny areas.

Edible Parts: The grains are edible at any stage of development. When young, the grains are milky and edible raw. Boil the older grains. Sorghum is a nutritious food.

Other Uses: Use the stems of tall sorghum as building materials.

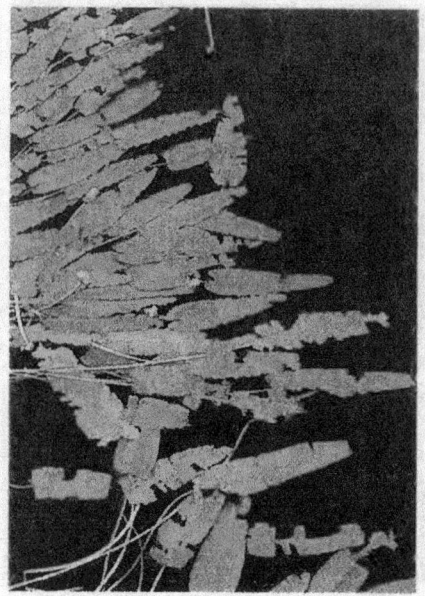 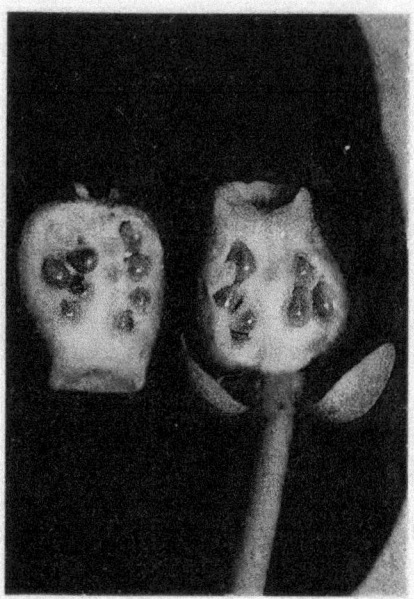

Spatterdock or yellow water lily
Nuphar species

Description: This plant has leaves up to 60 centimeters long with a triangular notch at the base. The shape of the leaves is somewhat variable. The plant's yellow flowers are 2.5 centimeters across and develop into bottle-shaped fruits. The fruits are green when ripe.

Habitat and Distribution: These plants grow throughout most of North America. They are found in quiet, fresh, shallow water (never deeper than 1.8 meters).

Edible Parts: All parts of the plant are edible. The fruits contain several dark brown seeds you can parch or roast and then grind into flour. The large rootstock contains starch. Dig it out of the mud, peel off the outside, and boil the flesh. Sometimes the rootstock contains large quantities of a very bitter compound. Boiling in several changes of water may remove the bitterness.

Sterculia
Sterculia foetida

Description: Sterculias are tall trees, rising in some instances to 30 meters. Their leaves are either undivided or palmately lobed. Their flowers are red or purple. The fruit of all sterculias is similar in aspect, with a red, segmented seedpod containing many edible black seeds.

Habitat and Distribution: There are over 100 species of sterculias distributed through all warm or tropical climates. They are mainly forest trees.

Edible Parts: The large, red pods produce a number of edible seeds. The seeds of all sterculias are edible and have a pleasant taste similar to cocoa. You can eat them like nuts, either raw or roasted.

CAUTION
Avoid eating large quantities. The seeds may have a laxative effect.

Strawberry
Fragaria species

Description: Strawberry is a small plant with a three-leaved growth pattern. It has small, white flowers usually produced during the spring. Its fruit is red and fleshy.

Habitat and Distribution: Strawberries are found in the North Temperate Zone and also in the high mountains of the southern Western Hemisphere. Strawberries prefer open, sunny areas. They are commonly planted.

Edible Parts: The fruit is edible fresh, cooked, or dried. Strawberries are a good source of vitamin C. You can also eat the plant's leaves or dry them and make a tea with them.

> **WARNING**
> Eat only white-flowering true strawberries. Other similar plants without white flowers can be poisonous.

Sugarcane
Saccharum officinarum

Description: This plant grows up to 4.5 meters tall. It is a grass and has grasslike leaves. Its green or reddish stems are swollen where the leaves grow. Cultivated sugarcane seldom flowers.

Habitat and Distribution: Look for sugarcane in fields. It grows only in the tropics (throughout the world). Because it is a crop, it is often found in large numbers.

Edible Parts: The stem is an excellent source of sugar and is very nutritious. Peel the outer portion off with your teeth and eat the sugarcane raw. You can also squeeze juice out of the sugarcane.

Sugar palm
Arenga pinnata

Description: This tree grows about 15 meters high and has huge leaves up to 6 meters long. Needlelike structures stick out of the bases of the leaves. Flowers grow below the leaves and form large conspicuous clusters from which the fruits grow.

Habitat and Distribution: This palm is native to the East Indies but has been planted in many parts of the tropics. It can be found at the margins of forests.

Edible Parts: The chief use of this palm is for sugar. However, its seeds and the tip of its stems are a survival food. Bruise a young flower stalk with a stone or similar object and collect the juice as it comes out. It is an excellent source of sugar. Boil the seeds. Use the tip of the stems as a vegetable.

CAUTION
The flesh covering the seeds may cause dermatitis.

Other Uses: The shaggy material at the base of the leaves makes an excellent rope as it is strong and resists decay.

Sweetsop
Annona squamosa

Description: This tree is small, seldom more than 6 meters tall, and multi-branched. It has alternate, simple, elongate, dark green leaves. Its fruit is green when ripe, round in shape, and covered with protruding bumps on its surface. The fruit's flesh is white and creamy.

Habitat and Distribution: Look for sweetsop at margins of fields, near villages, and around homesites in tropical regions.

Edible Parts: The fruit flesh is edible raw.

Other Uses: You can use the finely ground seeds as an insecticide.

CAUTION
The ground seeds are extremely dangerous to the eyes.

Tamarind
Tamarindus indica

Description: The tamarind is a large, densely branched tree, up to 25 meters tall. It has pinnate leaves (divided like a feather) with 10 to 15 pairs of leaflets.

Habitat and Distribution: The tamarind grows in the drier parts of Africa, Asia, and the Philippines. Although it is thought to be a native of Africa, it has been cultivated in India for so long that it looks like a native tree. It is also found in the American tropics, the West Indies, Central America, and tropical South America.

Edible Parts: The pulp surrounding the seeds is rich in vitamin C and is an important survival food. You can make a pleasantly acid drink by mixing the pulp with water and sugar or honey and letting the mixture mature for several days. Suck the pulp to relieve thirst. Cook the young, unripe fruits or seedpods with meat. Use the young leaves in soup. You must cook the seeds. Roast them above a fire or in ashes. Another way is to remove the seed coat and soak the seeds in salted water and grated coconut for 24 hours, then cook them. You can peel the tamarind bark and chew it.

Taro, cocoyam, elephant ears, eddo, dasheen
Colocasia and *Alocasia* species

Description: All plants in these groups have large leaves, sometimes up to 1.8 meters tall, that grow from a very short stem. The rootstock is thick and fleshy and filled with starch.

Habitat and Distribution: These plants grow in the humid tropics. Look for them in fields and near homesites and villages.

Edible Parts: All parts of the plant are edible when boiled or roasted. When boiling, change the water once to get rid of any poison.

CAUTION
If eaten raw, these plants will cause a serious inflammation of the mouth and throat.

Thistle
Cirsium species

Description: This plant may grow as high as 1.5 meters. Its leaves are long-pointed, deeply lobed, and prickly.

Habitat and Distribution: Thistles grow worldwide in dry woods and fields.

Edible Parts: Peel the stalks, cut them into short sections, and boil them before eating. The roots are edible raw or cooked.

CAUTION
Some thistle species are poisonous.

Other Uses: Twist the tough fibers of the stems to make a strong twine.

Ti
Cordyline terminalis

Description: The ti has unbranched stems with straplike leaves often clustered at the tip of the stem. The leaves vary in color and may be green or reddish. The flowers grow at the plant's top in large, plumelike clusters. The ti may grow up to 4.5 meters tall.

Habitat and Distribution: Look for this plant at the margins of forests or near homesites in tropical areas. It is native to the Far East but is now widely planted in tropical areas worldwide.

Edible Parts: The roots and very tender young leaves are good survival food. Boil or bake the short, stout roots found at the base of the plant. They are a valuable source of starch. Boil the very young leaves to eat. You can use the leaves to wrap other food to cook over coals or to steam.

Other Uses: Use the leaves to cover shelters or to make a rain cloak. Cut the leaves into liners for shoes; this works especially well if you have a blister. Fashion temporary sandals from the ti leaves. The terminal leaf, if not completely unfurled, can be used as a sterile bandage. Cut the leaves into strips, then braid the strips into rope.

Tree fern
Various genera

Description: Tree ferns are tall trees with long, slender trunks that often have a very rough, barklike covering. Large, lacy leaves uncoil from the top of the trunk.

Habitat and Distribution: Tree ferns are found in wet, tropical forests.

Edible Parts: The young leaves and the soft inner portion of the trunk are edible. Boil the young leaves and eat as greens. Eat the inner portion of the trunk raw or bake it.

Tropical Almond
Terminalia catappa

Description: This tree grows up to 9 meters tall. Its leaves are evergreen, leathery, 45 centimeters long, 15 centimeters wide, and very shiny. It has small, yellowish-green flowers. Its fruit is flat, 10 centimeters long, and not quite as wide. The fruit is green when ripe.

Habitat and Distribution: This tree is usually found growing near the ocean. It is a common and often abundant tree in the Caribbean and Central and South America. It is also found in the tropical rain forests of southeastern Asia, northern Australia, and Polynesia.

Edible Parts: The seed is a good source of food. Remove the fleshy, green covering and eat the seed raw or cooked.

 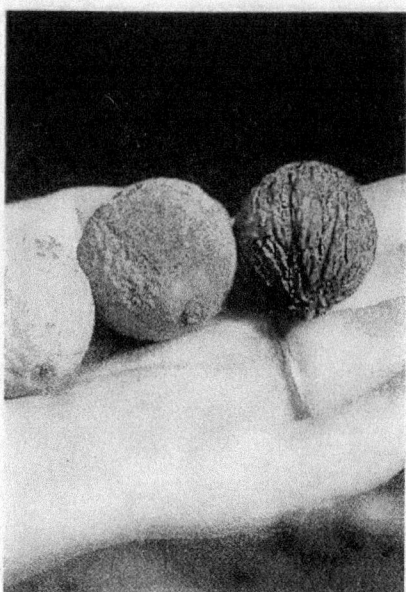

Walnut
Juglans species

Description: Walnuts grow on very large trees, often reaching 18 meters tall. The divided leaves characterize all walnut species. The walnut itself has a thick outer husk that must be removed to reach the hard inner shell of the nut.

Habitat and Distribution: The English walnut, in the wild state, is found from southeastern Europe across Asia to China and is abundant in the Himalayas. Several other species of walnut are found in China and Japan. The black walnut is common in the eastern United States.

Edible Parts: The nut kernel ripens in the autumn. You get the walnut meat by cracking the shell. Walnut meats are highly nutritious because of their protein and oil content.

Other Uses: You can boil walnuts and use the juice as an antifungal agent. The husks of "green" walnuts produce a dark brown dye for clothing or camouflage. Crush the husks of "green" black walnuts and sprinkle them into sluggish water or ponds for use as fish poison.

Water chestnut
Trapa natans

Description: The water chestnut is an aquatic plant that roots in the mud and has finely divided leaves that grow underwater. Its floating leaves are much larger and coarsely toothed. The fruits, borne underwater, have four sharp spines on them.

Habitat and Distribution: The water chestnut is a freshwater plant only. It is a native of Asia but has spread to many parts of the world in both temperate and tropical areas.

Edible Parts: The fruits are edible raw or cooked. The seeds are also a source of food.

Water lettuce
Ceratopteris species

Description: The leaves of water lettuce are much like lettuce and are very tender and succulent. One of the easiest ways of distinguishing water lettuce is by the little plantlets that grow from the margins of the leaves. These little plantlets grow in the shape of a rosette. Water lettuce plants often cover large areas in the regions where they are found.

Habitat and Distribution: Found in the tropics throughout the Old World in both Africa and Asia. Another kind is found in the New World tropics from Florida to South America. Water lettuce grows only in very wet places and often as a floating water plant. Look for water lettuce in still lakes, ponds, and the backwaters of rivers.

Edible Parts: Eat the fresh leaves like lettuce. Be careful not to dip the leaves in the contaminated water in which they are growing. Eat only the leaves that are well out of the water.

CAUTION
This plant has carcinogenic properties and should only be used as a last resort.

Water lily
Nymphaea odorata

Description: These plants have large, triangular leaves that float on the water's surface, large, fragrant flowers that are usually white or red, and thick, fleshy rhizomes that grow in the mud.

Habitat and Distribution: Water lilies are found throughout much of the temperate and subtropical regions.

Edible Parts: The flowers, seeds, and rhizomes are edible raw or cooked. To prepare rhizomes for eating, peel off the corky rind. Eat raw, or slice thinly, allow to dry, and then grind into flour. Dry, parch, and grind the seeds into flour.

Other Uses: Use the liquid resulting from boiling the thickened root in water as a medicine for diarrhea and as a gargle for sore throats.

Water plantain
Alisma plantago-aquatica

Description: This plant has small, white flowers and heart-shaped leaves with pointed tips. The leaves are clustered at the base of the plant.

Habitat and Distribution: Look for this plant in fresh water and in wet, full sun areas in Temperate and Tropical Zones.

Edible Parts: The rootstocks are a good source of starch. Boil or soak them in water to remove the bitter taste.

CAUTION
To avoid parasites, always cook aquatic plants.

Wild caper
Capparis aphylla

Description: This is a thorny shrub that loses its leaves during the dry season. Its stems are gray-green and its flowers pink.

Habitat and Distribution: These shrubs form large stands in scrub and thorn forests and in desert scrub and waste. They are common throughout North Africa and the Middle East.

Edible Parts: The fruits and the buds of young shoots are edible raw.

Wild crab apple or wild apple
Malus species

Description: Most wild apple look enough like domestic apples that the survivor can easily recognize them. Wild apple varieties are much smaller than cultivated kinds; the largest kinds usually do not exceed 5 to 7.5 centimeters in diameter, and most often less. They have small, alternate, simple leaves and often have thorns. Their flowers are white or pink and their fruits reddish or yellowish.

Habitat and Distribution: They are found in the savanna regions of the tropics. In temperate areas, wild apple varieties are found mainly in forested areas. Most frequently, they are found on the edge of woods or in fields. They are found throughout the Northern Hemisphere.

Edible Parts: Prepare wild apples for eating in the same manner as cultivated kinds. Eat them fresh, when ripe, or cooked. Should you need to store food, cut the apples into thin slices and dry them. They are a good source of vitamins.

CAUTION
Apple seeds contain cyanide compounds. Do not eat.

Wild desert gourd or colocynth
Citrullus colocynthis

Description: The wild desert gourd, a member of the watermelon family, produces a 2.4- to 3-meter-long ground-traveling vine. The perfectly round gourds are as large as an orange. They are yellow when ripe.

Habitat and Distribution: This creeping plant can be found in any climatic zone, generally in desert scrub and waste areas. It grows abundantly in the Sahara, in many Arab countries, on the southeastern coast of India, and on some of the islands of the Aegean Sea. The wild desert gourd will grow in the hottest localities.

Edible Parts: The seeds inside the ripe gourd are edible after they are completely separated from the very bitter pulp. Roast or boil the seeds—their kernels are rich in oil. The flowers are edible. The succulent stem tips can be chewed to obtain water.

Wild dock and wild sorrel
Rumex crispus and *Rumex acetosella*

Description: Wild dock is a stout plant with most of its leaves at the base of its stem that is commonly 15 to 30 centimeters long. The plants usually develop from a strong, fleshy, carrotlike taproot. Its flowers are usually very small, growing in green to purplish plumelike clusters. Wild sorrel is similar to the wild dock but smaller. Many of the basal leaves are arrow-shaped but smaller than those of the dock and contain a sour juice.

Habitat and Distribution: These plants can be found in almost all climatic zones of the world, in areas of high as well as low rainfall. Many kinds are found as weeds in fields, along roadsides, and in waste places.

Edible Parts: Because of the tender nature of the foliage, the sorrel and the dock are useful plants, especially in desert areas. You can eat their succulent leaves fresh or slightly cooked. To take away the strong taste, change the water once or twice during cooking. This latter tip is a useful hint in preparing many kinds of wild greens.

Wild fig
Ficus species

Description: These trees have alternate, simple leaves with entire margins. Often, the leaves are dark green and shiny. All figs have a milky, sticky juice. The fruits vary in size depending on the species, but are usually yellow-brown when ripe.

Habitat and Distribution: Figs are plants of the tropics and semitropics. They grow in several different habitats, including dense forests, margins of forests, and around human settlements.

Edible Parts: The fruits are edible raw or cooked. Some figs have little flavor.

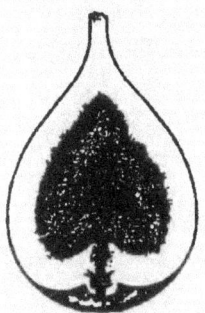

Wild gourd or luffa sponge
Luffa cylindrica

Description: The luffa sponge is widely distributed and fairly typical of a wild squash. There are several dozen kinds of wild squashes in tropical regions. Like most squashes, the luffa is a vine with leaves 7.5 to 20 centimeters across having 3 lobes. Some squashes have leaves twice this size. Luffa fruits are oblong or cylindrical, smooth, and many-seeded. Luffa flowers are bright yellow. The luffa fruit, when mature, is brown and resembles a cucumber.

Habitat and Distribution: A member of the squash family, which also includes the watermelon, cantaloupe, and cucumber, the luffa sponge is widely cultivated throughout the Tropical Zone. It may be found in a semi-wild state in old clearings and abandoned gardens in rainforests and semievergreen seasonal forests.

Edible Parts: You can boil the young green (half-ripe) fruits and eat them as a vegetable. Adding coconut milk will improve the flavor. After ripening, the luffa sponge develops an inedible spongelike texture in the interior of the fruit. You can also eat the tender shoots, flowers, and young leaves after cooking them. Roast the mature seeds a little and eat them like peanuts.

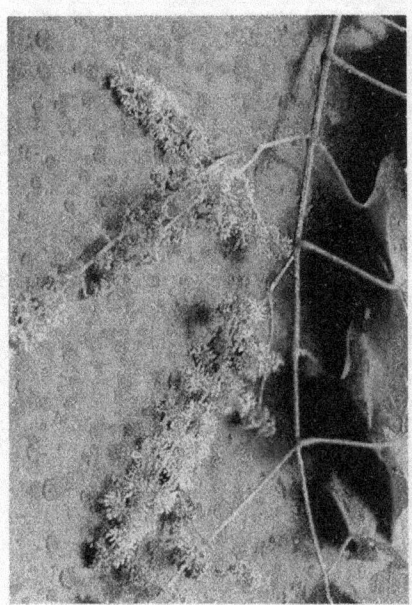

Wild grape vine
Vitis species

Description: The wild grape vine climbs with the aid of tendrils. Most grape vines produce deeply lobed leaves similar to the cultivated grape. Wild grapes grow in pyramidal, hanging bunches and are black-blue to amber, or white when ripe.

Habitat and Distribution: Wild grapes are distributed worldwide. Some kinds are found in deserts, others in temperate forests, and others in tropical areas. Wild grapes are commonly found throughout the eastern United States as well as in the southwestern desert areas. Most kinds are rampant climbers over other vegetation. The best place to look for wild grapes is on the edges of forested areas. Wild grapes are also found in Mexico. In the Old World, wild grapes are found from the Mediterranean region eastward through Asia, the East Indies, and to Australia. Africa also has several kinds of wild grapes.

Edible Parts: The ripe grape is the portion eaten. Grapes are rich in natural sugars and, for this reason, are much sought after as a source of energy-giving wild food. None are poisonous.

Other Uses: You can obtain water from severed grape vine stems. Cut off the vine at the bottom and place the cut end in a container. Make a slant-wise cut into the vine about 1.8 meters up on the hanging part. This cut will allow water to flow from the bottom end. As water diminishes in volume, make additional cuts further down the vine.

CAUTION
To avoid poisoning, do not eat grapelike fruits with only a *single* seed (moonseed).

 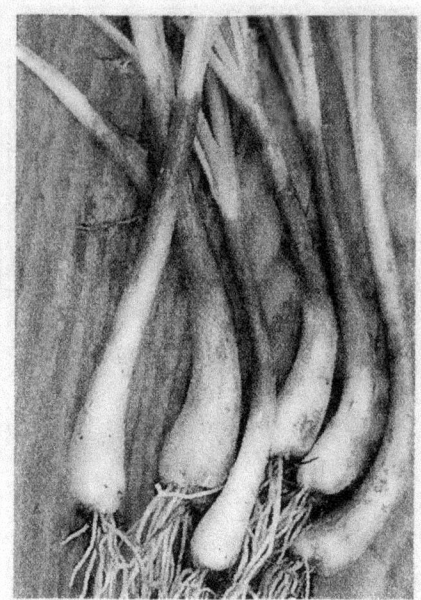

Wild onion and garlic
Allium species

Description: *Allium cernuum* is an example of the many species of wild onions and garlics, all easily recognized by their distinctive odor.

Habitat and Distribution: Wild onions and garlics are found in open, sunny areas throughout the temperate regions. Cultivated varieties are found anywhere in the world.

Edible Parts: The bulbs and young leaves are edible raw or cooked. Use in soup or to flavor meat.

CAUTION
There are several plants with onionlike bulbs that are extremely poisonous. Be certain that the plant you are using is a true onion or garlic. Do not eat bulbs with no onion smell.

Other Uses: Eating large quantities of onions will give your body an odor that will help to repel insects. Garlic juice works as an antibiotic on wounds.

Wild pistachio
Pistacia species

Description: Some kinds of pistachio trees are evergreen, while others lose their leaves during the dry season. The leaves alternate on the stem and have either three large leaves or a number of leaflets. The fruits or nuts are usually hard and dry at maturity.

Habitat and Distribution: About seven kinds of wild pistachio nuts are found in desert or semidesert areas surrounding the Mediterranean Sea to Turkey and Afghanistan. It is generally found in evergreen scrub forests or scrub and thorn forests.

Edible Parts: You can eat the oil nut kernels after parching them over coals.

Wild rice
Zizania aquatica

Description: Wild rice is a tall grass that averages 1 to 1.5 meters in height, but may reach 4.5 meters. Its grain grows in very loose heads at the top of the plant and is dark brown or blackish when ripe.

Habitat and Distribution: Wild rice grows only in very wet areas in tropical and temperate regions.

Edible Parts: During the spring and summer, the central portion of the lower stems and root shoots are edible. Remove the tough covering before eating. During the late summer and fall, collect the straw-covered husks. Dry and parch the husks, break them, and remove the rice. Boil or roast the rice and then beat it into flour.

Wild rose
Rosa species

Description: This shrub grows 60 centimeters to 2.5 meters high. It has alternate leaves and sharp prickles. Its flowers may be red, pink, or yellow. Its fruit, called rose hip, stays on the shrub year-round.

Habitat and Distribution: Look for wild roses in dry fields and open woods throughout the Northern Hemisphere.

Edible Parts: The flowers and buds are edible raw or boiled. In an emergency, you can peel and eat the young shoots. You can boil fresh, young leaves in water to make a tea. After the flower petals fall, eat the rose hips; the pulp is highly nutritious and an excellent source of vitamin C. Crush or grind dried rose hips to make flour.

CAUTION
Eat only the outer portion of the fruit as the seeds of some species are quite prickly and can cause internal distress.

Wood sorrel
Oxalis species

Description: Wood sorrel resembles shamrock or four-leaf clover, with a bell-shaped pink, yellow, or white flower.

Habitat and Distribution: Wood sorrel is found in Temperate Zones worldwide, in lawns, open areas, and sunny woods.

Edible Parts: Cook the entire plant.

CAUTION
Eat only small amounts of this plant as it contains a fairly high concentration of oxalic acid that can be harmful.

Yam
Dioscorea species

Description: These plants are vines that creep along the ground. They have alternate, heart- or arrow-shaped leaves. Their rootstock may be very large and weigh many kilograms.

Habitat and Distribution: True yams are restricted to tropical regions where they are an important food crop. Look for yams in fields, clearings, and abandoned gardens. They are found in rain forests, semievergreen seasonal forests, and scrub and thorn forests in the tropics. In warm temperate areas, they are found in seasonal hardwood or mixed hardwood-coniferous forests, as well as some mountainous areas.

Edible Parts: Boil the rootstock and eat it as a vegetable.

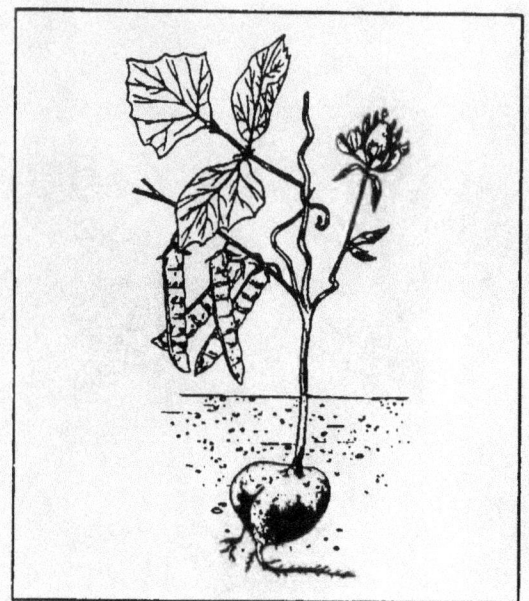
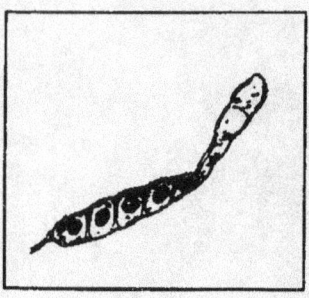

Yam bean
Pachyrhizus erosus

Description: The yam bean is a climbing plant of the bean family, with alternate, three-parted leaves and a turniplike root. The bluish or purplish flowers are pealike in shape. The plants are often so rampant that they cover the vegetation upon which they are growing.

Habitat and Distribution: The yam bean is native to the American tropics, but it was carried by man years ago to Asia and the Pacific islands. Now it is commonly cultivated in these places, and is also found growing wild in forested areas. This plants grows in wet areas of tropical regions.

Edible Parts: The tubers are about the size of a turnip and they are crisp, sweet, and juicy and have a nutty flavor. They are nourishing and at the same time quench the thirst. Eat them raw or boiled. To make flour, slice the raw tubers, let them dry in the sun, and grind into a flour that is high in starch and may be used to thicken soup.

CAUTION
The raw seeds are poisonous.

PART 2

POISONOUS PLANTS

Plants basically poison on contact, ingestion, or by absorption or inhalation. They cause painful skin irritations upon contact, they cause internal poisoning when eaten, and they poison through skin absorption or inhalation in respiratory system. Many edible plants have deadly relatives and look-alikes. Positive identification of edible plants will eliminate the danger of accidental poisoning. There is no room for experimentation where plants are concerned, especially in unfamiliar territory.

Plant poisoning ranges from minor irritation to death. A common question asked is, "How poisonous is this plant?" It is difficult to say how poisonous plants are because—

- Some plants require contact with a large amount of the plant before noticing any adverse reaction while others cause death with only a small amount.
- Every plant will vary in the amount of toxins it contains due to different growing conditions and slight variations in subspecies.
- Every person has a different level of resistance to toxic substances.
- Some persons may be more sensitive to a particular plant.

Some common misconceptions about poisonous plants are—

- *Watch the animals eat and what they eat.* Most of the time this statement is true, but some animals can eat plants that are poisonous to humans.
- *Boil the plant in water and any poisons will be removed.* Boiling removes many poisons, but not all.
- *Plants with a red color are poisonous.* Some plants that are red are poisonous, but not all.

Rules for Avoiding Poisonous Plants

Your best policy is to be able to look at a plant and identify it with absolute certainty and to know its uses or dangers. Many times this is not possible. If you have little or no knowledge of the local vegetation, use the rules to select plants for the "Universal Edibility Test" (see Appendix). Remember, avoid—

- *All mushrooms.* Mushroom identification is very difficult and must be precise, even more so than with other plants. Some mushrooms cause death very quickly. Some mushrooms have no known antidote. Two general types of mushroom poisoning are gastrointestinal and central nervous system.
- *Contact with or touching plants unnecessarily.*

Contact Dermatitis

Contact dermatitis from plants will usually cause the most trouble in the field. The effects may be persistent, spread by scratching, and are particularly dangerous if there is contact in or around the eyes.

The principal toxin of these plants is usually an oil that gets on the skin upon contact with the plant. The oil can also get on equipment and then infect whoever touches the equipment. Never burn a contact poisonous plant because the smoke may be as harmful as the plant. There is a greater danger of being affected when overheated and sweating. The infection may be local or it may spread over the body.

Symptoms may take from a few hours to several days to appear. Signs and symptoms can include burning, reddening, itching, swelling, and blisters.

When you first contact the poisonous plants or the first symptoms appear, try to remove the oil by washing with soap and cold water. If water is not available, wipe your skin repeatedly with dirt or sand. Do not use dirt if blisters have developed. The dirt may break open the blisters and leave the body open to infection. After you have removed the oil, dry the area. You can wash with a tannic acid solution and crush and rub jewelweed on the affected area to treat plant-caused rashes. You can make tannic acid from oak bark.

Ingestion Poisoning

Ingestion poisoning can be very serious and could lead to death very quickly. Do not eat any plant unless you have positively identified it first. Keep a log of all plants eaten.

Signs and symptoms of ingestion poisoning can include nausea, vomiting, diarrhea, abdominal cramps, depressed heartbeat and respiration, headaches, hallucinations, dry mouth, unconsciousness, coma, and death.

If you suspect plant poisoning, try to remove the poisonous material from the victim's mouth and stomach as soon as possible. Induce vomiting by tickling the back of his throat or by giving him some warm saltwater, if he is conscious. Dilute the poison by administering large quantities of water or milk, if he is conscious.

Castor bean, castor-oil plant, palma Christi
Ricinus communis
Spurge (*Euphorbiaceae*) Family

Description: The castor bean is a semiwoody plant with large, alternate, starlike leaves that grows as a tree in tropical regions and as an annual in temperate regions. Its flowers are very small and inconspicuous. Its fruits grow in clusters at the tops of the plants.

CAUTION
All parts of the plant are very poisonous to eat. The seeds are large and may be mistaken for a beanlike food.

Habitat and Distribution: This plant is found in all tropical regions and has been introduced to temperate regions.

Chinaberry
Melia azedarach
Mahogany (*Meliaceae*) Family

Description: This tree has a spreading crown and grows up to 14 meters tall. It has alternate, compound leaves with toothed leaflets. Its flowers are light purple with a dark center and grow in ball-like masses. It has marble-sized fruits that are light orange when first formed but turn lighter as they become older.

CAUTION
All parts of the tree should be considered dangerous if eaten. Its leaves are a natural insecticide and will repel insects from stored fruits and grains. Take care not to eat leaves mixed with the stored food.

Habitat and Distribution: Chinaberry is native to the Himalayas and eastern Asia but is now planted as an ornamental tree throughout the tropical and subtropical regions. It has been introduced to the southern United States and has escaped to thickets, old fields, and disturbed areas.

Cowhage, cowage, cowitch
Mucuna pruritum
Leguminosae (*Fabaceae*) Family

Description: A vinelike plant that has oval leaflets in groups of three and hairy spikes with dull purplish flowers. The seeds are brown, hairy pods.

CAUTION
Contact with the pods and flowers causes irritation and blindness if in the eyes.

Habitat and Distribution: Tropical areas and the United States.

Death camas, death lily
Zigadenus species
Lily (*Liliaceae*) Family

Description: This plant arises from a bulb and may be mistaken for an onionlike plant. Its leaves are grasslike. Its flowers are six-parted and the petals have a green, heart-shaped structure on them. The flowers grow on showy stalks above the leaves.

CAUTION
All parts of this plant are very poisonous. Death camas does not have the onion smell.

Habitat and Distribution: Death camas is found in wet, open, sunny habitats, although some species favor dry, rocky slopes. They are common in parts of the western United States. Some species are found in the eastern United States and in parts of the North American western subarctic and eastern Siberia.

Lantana

Lantana camara
Vervain (*Verbenaceae*) Family

Description: Lantana is a shrublike plant that may grow up to 45 centimeters high. It has opposite, round leaves and flowers borne in flat-topped clusters. The flower color (which varies in different areas) may be white, yellow, orange, pink, or red. It has a dark blue or black berrylike fruit. A distinctive feature of all parts of this plant is its strong scent.

> **CAUTION**
> All parts of this plant are poisonous if eaten and can be fatal. This plant causes dermatitis in some individuals.

Habitat and Distribution: Lantana is grown as an ornamental in tropical and temperate areas and has escaped cultivation as a weed along roads and old fields.

Manchineel
Hippomane mancinella
Spurge (*Euphorbiaceae*) Family

Description: Manchineel is a tree reaching up to 15 meters high with alternate, shiny green leaves and spikes of small greenish flowers. Its fruits are green or greenish-yellow when ripe.

CAUTION
This tree is extremely toxic. It causes severe dermatitis in most individuals after only .5 hour. Even water dripping from the leaves may cause dermatitis. The smoke from burning it irritates the eyes. No part of this plant should be considered a food.

Habitat and Distribution: This tree prefers coastal regions. Found in south Florida, the Caribbean, Central America, and northern South America.

Oleander
Nerium oleander
Dogbane (*Apocynaceae*) Family

Description: This shrub or small tree grows to about 9 meters, with alternate, very straight, dark green leaves. Its flowers may be white, yellow, red, pink, or intermediate colors. Its fruit is a brown, podlike structure with many small seeds.

CAUTION
All parts of the plant are very poisonous. Do not use the wood for cooking; it gives off poisonous fumes that can poison food.

Habitat and Distribution: This native of the Mediterranean area is now grown as an ornamental in tropical and temperate regions.

Pangi
Pangium edule
Pangi Family

Description: This tree, with heart-shaped leaves in spirals, reaches a height of 18 meters. Its flowers grow in spikes and are green in color. Its large, brownish, pear-shaped fruits grow in clusters.

CAUTION
All parts are poisonous, especially the fruit.

Habitat and Distribution: Pangi trees grow in southeast Asia.

 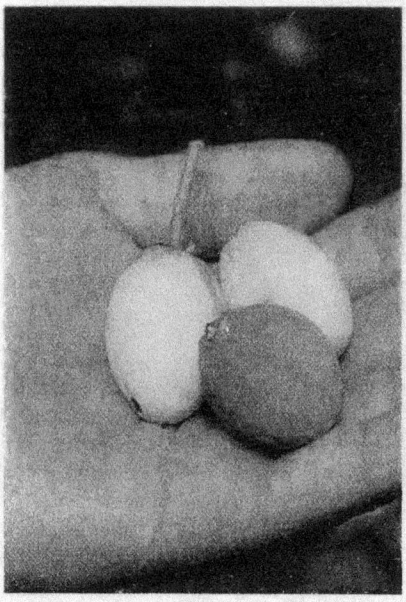

Physic nut
Jatropha curcas
Spurge (*Euphorbiaceae*) Family

Description: This shrub or small tree has large, 3- to 5-parted alternate leaves. It has small, greenish-yellow flowers and its yellow, apple-sized fruits contain three large seeds.

CAUTION
The seeds taste sweet but their oil is violently purgative. All parts of the physic nut are poisonous.

Habitat and Distribution: Throughout the tropics and southern United States.

Poison hemlock, fool's parsley
Conium maculatum
Parsley (*Apiaceae*) Family

Description: This biennial herb may grow to 2.5 meters high. The smooth, hollow stem may or may not be purple or red striped or mottled. Its white flowers are small and grow in small groups that tend to form flat umbels. Its long, turniplike taproot is solid.

CAUTION
This plant is very poisonous and even a very small amount may cause death. This plant is easy to confuse with wild carrot or Queen Anne's lace, especially in its first stage of growth. Wild carrot or Queen Anne's lace has hairy leaves and stems and smells like carrot. Poison hemlock does not.

Habitat and Distribution: Poison hemlock grows in wet or moist ground like swamps, wet meadows, stream banks, and ditches. Native to Eurasia, it has been introduced to the United States and Canada.

Poison ivy and poison oak
Toxicodendron radicans and *Toxicodendron diversibba*
Cashew (*Anacardiaceae*) Family

Description: These two plants are quite similar in appearance and will often crossbreed to make a hybrid. Both have alternate, compound laves with three leaflets. The leaves of poison ivy are smooth or serrated. Poison oak's leaves are lobed and resemble oak leaves. Poison ivy grows as a vine along the ground or climbs by red feeder roots. Poison oak grows like a bush. The greenish-white flowers are small and inconspicuous and are followed by waxy green berries that turn waxy white or yellow, then gray.

CAUTION
All parts, at all times of the year, can cause serious contact dermatitis.

Habitat and Distribution: Poison ivy and oak can be found in almost any habitat in North America.

Poison sumac
Toxicodendron vernix
Cashew (*Anacardiaceae*) Family

Description: Poison sumac is a shrub that grows to 8.5 meters tall. It has alternate, pinnately compound leafstalks with 7 to 13 leaflets. Flowers are greenish-yellow and inconspicuous and are followed by white or pale yellow berries.

CAUTION
All parts can cause serious contact dermatitis at all times of the year.

Habitat and Distribution: Poison sumac grows only in wet, acid swamps in North America.

Renghas tree, rengas tree, marking nut, black-varnish tree
Gluta
Cashew (*Anacardiaceae*) Family

Description: This family comprises about 48 species of trees or shrubs with alternating leaves in terminal or axillary panicles. Flowers are similar to those of poison ivy and oak.

> **CAUTION**
> Can cause contact dermatitis similar to poison ivy or poison oak.

Habitat and Distribution: India, east to Southeast Asia.

 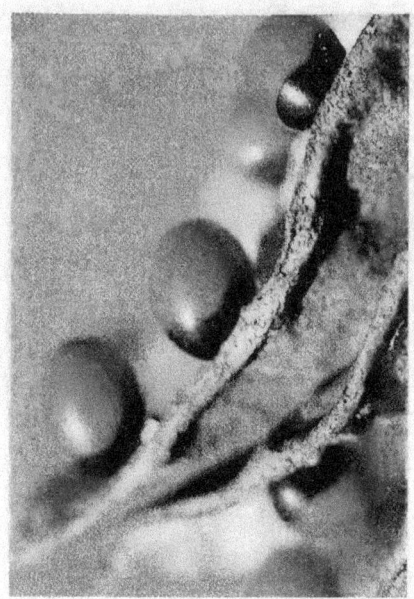

Rosary pea or crab's eyes
Abrus precatorius
Leguminosae (*Fabaceae*) Family

Description: This plant is a vine with alternate compound leaves, light purple flowers, and beautiful seeds that are red and black.

CAUTION
This plant is one of the most dangerous plants. One seed may contain enough poison to kill an adult.

Habitat and Distribution: This is a common weed in parts of Africa, southern Florida, Hawaii, Guam, the Caribbean, and Central and South America.

Strychnine tree
Nux vomica
Logania (*Loganiaceae*) Family

Description: The strychnine tree is a medium-sized evergreen, reaching a height of about 12 meters, with a thick, frequently crooked trunk. Its deeply veined oval leaves grow in alternate pairs. Small, loose clusters of greenish flowers appear at the ends of branches and are followed by fleshy, orange-red berries about 4 centimeters in diameter.

CAUTION
The berries contain the disklike seeds that yield the poisonous substance strychnine. All parts of the plant are poisonous.

Habitat and Distribution: A native of the tropics and subtropics of southeastern Asia and Australia.

Trumpet vine or trumpet creeper
Campsis radicans
Trumpet creeper (*Bignoniaceae*) Family

Description: This woody vine may climb to 15 meters high. It has pealike fruit capsules. The leaves are pinnately compound, 7 to 11 toothed leaves per leaf stock. The trumpet-shaped flowers are orange to scarlet in color.

CAUTION
This plant causes contact dermatitis.

Habitat and Distribution: This vine is found in wet woods and thickets throughout eastern and central North America.

Water hemlock or spotted cowbane
Cicuta maculata
Parsley (*Apiaceae*) Family

Description: This perennial herb may grow to 1.8 meters high. The stem is hollow and sectioned off like bamboo. It may or may not be purple or red striped or mottled. Its flowers are small, white, and grow in groups that tend to form flat umbels. Its roots may have hollow air chambers and, when cut, may produce drops of yellow oil.

CAUTION
This plant is very poisonous and even a very small amount of this plant may cause death. Its roots have been mistaken for parsnips.

Habitat and Distribution: Water hemlock grows in wet or moist ground like swamps, wet meadows, stream banks, and ditches throughout the United States and Canada.

INDEX

A

Abal *(Calligonum comosum)* 7
Abrasive, plant for 36
Abrus precatorius *(Rosary pea)*..135
Absorption poisoning.................119
Acacia *(Acacia farnesiana)*..............8
Adansonia digitata (Baobab)........18
Aegle marmelos (Bael fruit) 15
Agave *(Agave* species) 9
Alisma plantago-aquatica
   (Water plantain)...................103
Allium species (Wild onion) 111
Almond *(Prunus amygdalus)*........10
Alocasia species (Elephant ears)...94
Amaranth *(Amaranthus* species)..11
Anacardium occidentale
   (Cashew nut) 30
Animals, and poisonous plants..120
Annona squamosa (Sweetsop) 92
Antibiotic, plants as 111
Antidesma bunius (Bignay)...........22
Antifungal agent, plant as.............99
Arctium lappa (Burdock)..............26
Arctostaphylos uvaursi (Bearberry)20
Arenga pinnata (Sugar palm) 91
Arrowroot *(Maranta* and
   Sagittaria species) 13
Artic Willow *(Salix arctica)*..........12
Artocarpus incisa (Breadfruit) 25
Asparagus *(Asparagus officinalis)*..14
Atriplex halimus (Sea orach) 84
Atriplex species (Orach) 65

B

Bael fruit *(Aegle marmelos)* 15
Bamboo *(Bambusa, Dendrocalamus,*
   Phyllostachys) 26
Bambusa (Bamboo) 16
Banana *(Musa* species) 17
Bandages, plants for
   coconut 36
   ti 96
Baobab *(Adansonia digitata)*........18
Batoko plum
   (Flacourtia inermis) 19
Bearberry *(Arctostaphylos
   uvaursi)* 20
Beech *(Fagus* species)....................21
Bignay *(Antidesma bunius)*...........22
Birdlime, plant as..........................25
Black-varnish tree *(Gluta)*..........134
Blackberry *(Rubus* species)...........23
Blueberry *(Vaccinium* species) 24
Boiling, poisonous plants...........120
Breadfruit *(Artocarpus incisa)* 25
Building materials, plants for
   bamboo.....................................16
   coconut 36
   date palm..................................42
   sorghum....................................86
Bulb..2, 4
Burdock *(Arctium lappa)*..............26
Burl Palm *(Corypha elata)*............27

C

Cavanillesia platanifolia
  (Cuipo tree) 40
Calamus species (Rattan palm) 75
Calcium, plants high in 41
Calligonum comosum (Abal) 7
Caltha palustris
  (Marsh marigold) 60
Campsis radicans
  (Trumpet vine) 137
Canna lily *(Canna indica)* 28
Capparis aphylla (Wild caper) ... 104
Carcinogens, plants with 101
Carica papaya (Papaya) 67
Carob tree *(Ceratonia siliqua)* 29
Caryota urens (Fishtail palm) 47
Cashew *(Anacardiacese)*
  family 132, 133, 134
Cashew nut *(Anacardium*
  occidentale) 30
Castanea sativa (Chestnut) 33
Castor bean *(Ricinus*
  communis) 122
Castor-oil plant
  (Ricinus communis) 122
Cattail *(Typha latifolia)* 31
Caulk, plants for 25
Cedars 55
Celtis species (Hackberry) 50
Ceratonia siliqua (Carob tree) 29
Ceratopteris species
  (Water lettuce) 101
Cereus cactus *(Cereus* species) 32

Cetraria islandica
  (Iceland moss) 53
Chestnut *(Castanea sativa)* 33
Chicory *(Cichorium intybus)* 34
Chinaberry *(Melia azedarach)* 123
Chufa *(Cyperus esculentus)* 35
Cichorium intybus (Chicory) 34
Cicuta maculata
  (Water hemlock) 138
Cirsium species (Thistle) 95
Citrullus colocynthis
  (Wild desert gourd) 106
Cladonia rangifera
  (Reindeer moss) 77
Claytonia species
  (Indian potato) 54
Cleaning, plants for 42
Clove 2, 4
Coconut *(Cocos nucifera)* 36
Cocoyam *(Colocasia* species) 94
Coffee, substitutes for
  beech 21
  chicory 34
  chufa 35
  dandelion 41
  juniper 55
  oak 64
Colocasia species (Taro) 94
Colocynth *(Citrullus*
  colocynthis) 106
Color, of poisonous plants 120
Common jujube *(Ziziphus*
  jujuba) 37
Conium maculatum
  (Poison hemlock) 131

Contact dermatitis, plants
    causing 120–21
  agave 9
  burl palm 27
  lantana 126
  manchineel 127
  poison ivy/oak 132
  poison sumac 133
  renghas tree 134
  sugar palm 91
  trumpet vine 137
Contact poisoning 119, 120–21
Containers, plants for
  bamboo 16
  coconut 36
  rattan palm 75
Cordyline terminalis (Ti) 96
Corm 2–3, 4
Corrosion, plants protecting
    against 36
Corylus species
    (Hazelnut/wild filbert) 51
Corypha elata (Burl Palm) 27
Cowhage *(Mucuna pruritum)* 124
Cowitch *(Mucuna pruritum)*, ... 124
Crab's eyes *(Abrus precatorius)* .. 135
Cranberry *(Vaccinium
    macrocarpon)* 38
Crowberry *(Empetrum
    nigrum)* 39
Crown 2–3, 4
Cuipo tree *(Caavnillesia
    platanifolia)* 40
Cyanide, plants containing 105
Cyperus esculentus (Chufa) 35

D

Dandelion *(Taraxacum
    officinale)* 41
Dasheen (*Colocasia* species) 94
Date Palm *(Phoenix
    dactylifera)* 42
Daylily *(Hemerocallis fulva)* 43
Death camas
    (*Zigadenus* species) 125
Death lily (*Zigadenus* species) 125
Dendrocalamus (Bamboo) 16
Dermatitis. *See* Contact dermatitis,
    plants causing
Dewberry (*Rubus* species) 23
Diarrhea, plants causing
  asparagus 14
  daylily 43
Diarrhea, plants for treating
  baobab 18
  blackberry, raspberry, dewberry ... 23
  plantain 71
  water lily 102
Dioscorea species (Yam) 116
Diospyros virginiana
    (Persimmon) 68
Diuretic, plants acting as
  burdock 26
  cranberry 38
Dogbane *(Apocynaceae)* family .. 128
Duchesnea *(Duchesnea indica)* 44
Dyes, plants for
  pokeweed 72
  walnut 99

E

Eddo (*Colocasia* species)................94
Edibility, Universal Test for........5, 6
Elderberry (*Sambucus
    canadensis*)..............................45
Elephant ears (*Alocasia species*)....94
Empetrum nigrum (Crowberry)...39
Epilobium angustifolium
    (Fireweed)................................46
Eskimo potato
    (*Claytonia* species)..................54
Eugenia jambos (Rose apple)........79
Eye irritation, plants causing......120
    cowage...*124*
    papaya...*67*
    sweetsop.......................................*92*

F

Fagus species (Beech).....................21
Ficus species (Wild fig)...............108
Fire, and bamboo...........................16
Fireweed (*Epilobium
    angustifolium*).........................46
Fish poison, plants for..................99
Fish traps, plants for......................75
Fishtail palm (*Caryota urens*)........47
Flacourtia inermis
    (Batoko plum).........................19
Fool's parsley (*Conium
    maculatum*)...........................131
Foxtail grass (*Setaria* species)........48
Fragaria species (Strawberry)........89

G

Gaylussacia species
    (Huckleberry).........................24
Glue, plants for
    breadfruit....................................*25*
    dandelion....................................*41*
    pine...*70*
Gluta (Renghas tree)...................134
Goa Bean (*Psophocarpus
    tetragonolobus*).......................49

H

Hackberry (*Celtis* species)............50
Hallucinogenic plants...................61
Haloxylon ammondendron
    (Saxual)...................................82
Hazelnut (*Corylus* species)...........51
Hemerocallis fulva (Daylily)..........43
Hippomane mancinella
    (Manchineel).......................127
Hooks, plants for.............................9
Horseradish tree (*Moringa
    pterygosperma*).......................52
Huckleberry (*Gaylussacia*
    species)....................................24

I

Iceland moss (*Cetraria
    islandica*)................................53
Identification, plant........2, 2–3, 3, 4
Illness, plants causing. See also
    Plants, poisonous
    daylily..*43*

mulberry..61
oak..64
Illness, plants for treating
   baobab..18
   blackberry, raspberry, dewberry...23
   cranberry..38
   water lily.......................................102
Indian potato
   (*Claytonia* species)..................54
Indian strawberry
   (*Duchesnea indica*)................44
Ingestion poisoning............119, 121
Inhalation poisoning...................119
Insect repellent, plants for
   cattail..31
   chinaberry....................................123
   coconut...36
   sweetsop...92
   wild onion/garlic........................111
Insulation, plants for.....................31

J

Jatropha curcas (Physic nut).......130
Jewelweed, and dermatitis...........121
Juglans species (Walnut)...............99
Juniper (*Juniperus* species)...........55

K

Kidney failure, plants causing......64
Kinnikinnick (*Arctostaphylos*
   uvaursi)......................................20

L

Lantana (*Lantana camara*).........126
Laportea species (Nettle)..............62
Laxative effect, plants having
   bignay...22
   cereus cactus..................................32
   mulberry..61
   plantain..71
   sterculia...88
Leaves, types of....................2, **2, 3, 4**
Leguminosae *(Fabaceae)*
   family...............................124, 135
Lily (*Liliaceae*) family.................125
Lips, plants irritating....................30
Liquids, plants producing. *See also*
   Water, plants as source of
   agave...9
   burl palm.......................................27
   coconut..36
   fishtail palm.................................47
   nipa palm......................................63
   sugar palm....................................91
Logania (*Loganiaceae*) family.....136
Lotus (*Nelumbo* species)...............56
Luffa cylindrical (Wild gourd)....109
Luffa sponge (*Luffa cylindrical*).109

M

Mahogany *(Meliaceae)* family....123
Malanga (*Xanthosoma caracu*).....57
Malus species (Wild apple).........105
Mammillaria species
   (Pincushion cactus)................69

Manchineel *(Hippomane mancinella)*............................127
Mango *(Mangifera indica)*.............58
Manioc *(Manihot utillissima)*.......59
Maranta species (Arrowroot).......13
Marking nut *(Gluta)*....................134
Marsh marigold
    (Caltha palustris).....................60
Meat, plants tenderizing..............67
Melia azedarach (Chinaberry)....123
Metroxylon sagu (Sago palm).......80
Moringa pterygosperma
    (Horseradish tree).................52
Morus species (Mulberry)............61
Mouth, inflammation, plants
    causing....................................94
Mucuna pruritum (Cowhage)....124
Mulberry (*Morus* species)............61
Musa species.................................17
Mushrooms, avoiding.................120

N

Nausea, plants causing. *See also*
    Plants, poisonous
    asparagus................................*14*
    mulberry.................................*61*
Nelumbo species (Lotus)...............56
Nerium oleander (Oleander) 128
Nettle (*Urtica* and
    Laportea species)....................62
Nipa palm *(Nipa fruticans)*..........63
Nitrogen, plants high in 7
Nuphar species (Spatterdock)......87

Nutrition, plants high in. *See also*
    Oil, plants high in; Sugar, plants
    high in; Vitamins, plants high in
    almond....................................*10*
    chestnut..................................*33*
    coconut...................................*36*
    mango......................................*58*
    nettle.......................................*62*
Nux vomica (Strychine tree).......136
Nymphaea odorata (Water lily)..102

O

Oak (*Quercus* species)...................64
Oak bark, and tannic acid..........121
Oil, of poisonous plants.......120–21
Oil, plants high in
    beech.......................................*21*
    coconut...................................*35*
    hazelnut..................................*51*
    walnut.....................................*99*
    wild desert gourd.................*106*
Oleander (*Nerium oleander*).....128
Opuntia species (Prickley pear
    cactus)....................................73
Orach (*Atriplex* species)...............65
Oxalic acid, plants
    containing......................85, 115
Oxalis species (Wood sorrel) 115

P

Pachyrhizus erosus (Yam bean)..117
Pain, plants for relieving.............71

Palma Christi
  (Ricinus communis) 122
Palmetto palm *(Sabal
  palmetto)* 66
Pandanus species (Screw pine)..... 83
Pangi *(Pangium edule)* 129
Papaya *(Carica papaya)* 67
Parsley *(Apiaceae)* family 131, 138
Parsnips, and water hemlock 138
Pawpaw *(Carica papaya)* 67
Persimmon *(Diospyros
  virginiana)* 68
Phoenix dactylifera (Date Palm) ... 42
Phragmites australis (Reed) 76
Phyllostachys (Bamboo) 16
Physic nut *(Jatropha curcas)* 130
Phytolacca americana
  (Pokeweed) 72
Pincushion cactus *(Mammillaria*
  species) 69
Pine *(Pinus* species) 70
Pistacia species (Wild
  pistachio) 112
Plantago species (Plantain) 71
Plantain *(Musa* species) 17
Plantain *(Plantago* species) 71
Plants
  identification of 2, 2–3, **3, 4**
  Universal Edibility Test 5, **6**
Plants, poisonous 119–38
  apple seeds 105
  asparagus fruit 14
  avoiding 120
  cashew nut hull 30

contact dermatitis 120–21
elderberry 45
grape-like fruit 110
ingestion poisoning 121
misconceptions about 120
onion-like bulbs 111
pokeweed 72
rock tripe 78
strawberry-like plants 89
thistle 95
types of 122–38
yam bean seeds 117
Poison hemlock
  (Conium maculatum) 131
Poison ivy *(Toxicodendron
  radicans)* 132
Poison oak *(Toxicodendron
  diversibba)* 132
Poison sumac *(Toxicodendron
  vernix)* 133
Poisonous plants. See Plants,
  poisonous
Pokeweed *(Phytolacca
  americana)* 72
Portulaca oleracea (Purslane) 74
Prickley pear cactus
  (Opuntia species) 73
Protein, plants containing
  goa bean 49
  walnut 99
Prunus amygdalus (Almond) 10
Psophocarpus tetragonolobus
  (Goa Bean) 49
Purslane *(Portulaca oleracea)* 74

Q

Queen Anne's lace, and
    poison hemlock 131
Quercus species (Oak) 64

R

Raspberry (*Rubus* species) 23
Rattan palm (*Calamus* species) 75
Reed *(Phragmites australis)* 76
Reindeer moss
    (Cladonia rangifera) 77
Rengas/Rhengas tree *(Gluta)* 134
Resin, uses of 70
Rhizome 2–3, 4
Rhubarb, and burdock 26
Ricinus communis (Castor bean) 122
Rock tripe (*Umbilicaria* species) .. 78
Roots, types of 2, 4
Rope, plants for making
    agave 9
    baobab 18
    cuipo tree 40
    sugar palm 91
    ti 96
Rosa species (Wild rose) 114
Rosary pea *(Abrus precatorius)* ... 135
Rose apple *(Eugenia jambos)* 79
Rubus species (Blackberry,
    raspberry) 23
Rumex acetosella (Sheep sorrel) ... 85
Rumex acetosella (Wild sorrel) ... 107
Rumex crispus (Wild dock) 107

S

Sabal palmetto (Palmetto palm) .. 66
Saccharum officinarum
    (Sugarcane) 90
Sagittaria species (Arrowroot) 13
Sago palm *(Metroxylon sagu)* 80
Saint John's bread 29
Salix arctica (Arctic willow) 12
Sambucus canadensis
    (Elderberry) 45
Sassafras *(Sassafras albidum)* 81
Saxual *(Haloxylon*
    ammondendron) 82
Scent, plant
    lantana 126
    wild onion/garlic 111
Screw pine (*Pandanus* species) 83
Sea orach *(Atriplex halimus)* 84
Seasoning, plants for
    horseradish tree 52
    juniper 55
Setaria species (Foxtail grass) 48
Sewing, plant for 9
Sheep sorrel *(Rumex acetosella)* ... 85
Shoes, plants for 96
Skin irritation, plants causing. *See*
    Contact dermatitis, plants
    causing
Skin, plants for treating 36
Soap, plants for 9
Sorghum (*Sorghum* species) 86
Spatterdock (*Nuphar* species) 87
Spotted cowbane
    (*Cicuta maculata*) 138

Spurge *(Euphorbiaceae)*
  family 122, 127, 130
Starch, plants high in
  arrowroot *13*
  canna lily *28*
  cattail *31*
  fishtail palm *47*
  lotus *56*
  malanga *57*
  manioc *59*
  rattan palm *75*
  sago palm *80*
  ti *96*
  water plantain *103*
Sterculia *(Sterculia foetida)* 88
Storage, plants used for
  banana and plantain *17*
  coconut *36*
Strawberry *(Fragaria* species) 89
Strychine tree *(Nux vomica)* 136
Sugar palm *(Arenga pinnata)* 91
Sugar, plants high in
  abal *7*
  coconut *36*
  nipa palm *63*
  pine *70*
  sugar palm *91*
  sugarcane *90*
  wild grape vine *110*
Sugarcane *(Saccharum officinarum)* 90
Sweat, plants inducing 26
Sweetsop *(Annona squamosa)* 92

Symptoms, of ingestion
  poisoning 121

T

Tamarind *(Tamarindus indica)* 93
Tannic acid
  and acorns 64
  and dermatitis 121
Taproot 2, 4
Taraxacum officinale
  (Dandelion) 41
Taro *(Colocasia* species) 94
Teeth, plants with uses for
  pine *70*
  sassafras *81*
Terminalia catappa
  (Tropical Almond) 98
Thatch, plants for
  coconut *36*
  date palm *42*
  nipa palm *63*
  sago palm *80*
Thistle *(Cirsium* species) 95
Throat
  plants causing inflammation 94
  plants for treating 102
Ti *(Cordyline terminalis)* 96
Tinder, plants for
  cattail *31*
  coconut *36*
Tongue, plants irritating 30
Tools, plants for making 16
Toxicodendron diversibba
  (Poison oak) 132

Toxicodendron radicans
   (Poison ivy) 132
Toxicodendron vernix
   (Poison sumac) 133
Trapa natans (Water chestnut) .. 100
Tree fern 97
Tropical Almond
   (*Terminalia catappa*) 98
Trumpet creeper
   (*Bignoniaceae*) family 137
Trumpet vine *(Campsis
   radicans)* 137
Tuber 2–3, 4
Twine, plants for
   mulberry 61
   nettle 62
   thistle 95
Typha latifolia (Cattail) 31

U

Umbilicaria species (Rock
   tripe) 78
Univeral Edibility Test 5, 6
Urinary infections, plants
   for treating 38
Urtica species (Nettle) 62
Utensils, plants for making 16

V

Vaccinium macrocarpon
   (Cranberry) 38
Vaccinium species (Blueberry) 24
Vervain *(Verbenaceae)* family 126

Vitamins, plants high in
   artic willow *12*
   bael fruit *15*
   common jujube *37*
   dandelion *41*
   papaya *67*
   persimmon *68*
   pine *70*
   strawberry *89*
   tamarind *93*
   wild apple *105*
   wild rose *114*
Vitis species (Wild grape vine) ... 110
Vomiting, inducing 121

W

Walnut (*Juglans* species) 99
Water chestnut *(Trapa natans)* .. 100
Water hemlock
   (*Cicuta maculata*) 138
Water lettuce (*Ceratopteris*
   species) 101
Water lily *(Nymphaea
   odorata)* 102
Water plantain *(Alisma
   plantago-aquatica)* 103
Water plants, danger of 60, 103
Water, plants as source of
   banana *17*
   baobab *18*
   cereus cactus *32*
   cuipo tree *40*
   nipa palm *63*
   pincushion cactus *69*

prickly pear cactus......................73
rattan palm 75
saxual 82
wild desert gourd......................106
wild grape vine..........................110
Waterproofing, plants for.............96
Weapons, plants for making........16
Weaving, plants for
    agave....................................9
    burdock...............................26
    burl palm............................27
    cattail..................................31
    coconut 36
    date palm...........................42
    nipa palm 63
Wild apple (*Malus* species) 105
Wild caper (*Capparis aphylla*) ...104
Wild carrot, and
    poison hemlock....................131
Wild crab apple
    (*Malus* species) 105
Wild desert gourd (*Citrullus
    colocynthis*)............................106
Wild dock (*Rumex crispus*) 107
Wild fig (*Ficus* species) 108
Wild filbert (*Corylus* species).......51
Wild garlic (*Allium* species) 111
Wild gourd (*Luffa cylindrical*)....109
Wild grape vine (*Vitis* species)...110

Wild onion (*Allium* species) 111
Wild pistachio
    (*Pistacia* species)..................112
Wild rice (*Zizania aquatica*) 113
Wild rose (*Rosa* species).............114
Wild sorrel (*Rumex acetosella*)...107
Wood sorrel (*Oxalis* species) 115
Wounds, plants for treating
    plantain 71
    prickly pear cactus......................73
    wild garlic..................................111

X

Xanthosoma caracu (Malanga) 57

Y

Yam (*Dioscorea* species) 116
Yam bean (*Pachyrhizus erosus*) ..117
Yellow water lily
    (*Nuphar* species).....................87

Z

Zigadenus species
    (Death camus) 125
Zizania aquatica (Wild rice) 113
Ziziphus jujuba
    (Common jujube) 37

www.ingramcontent.com/pod-product-compliance
Lightning Source LLC
Chambersburg PA
CBHW031357040426
42444CB00005B/325